AF335553

ADVANCES IN BIOPHYSICAL CHEMISTRY

Volume 2 • 1992

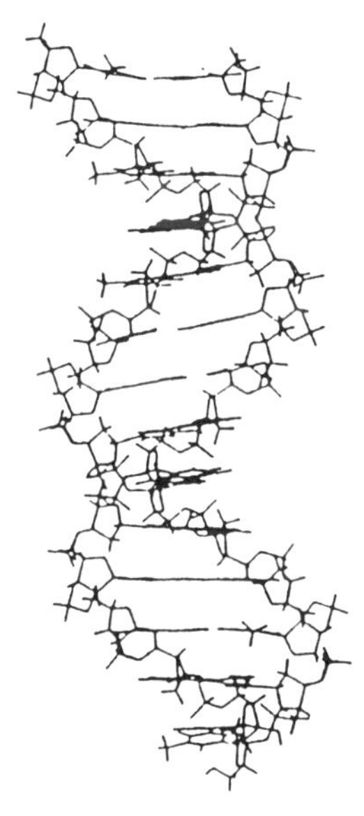

ADVANCES IN BIOPHYSICAL CHEMISTRY

A Research Annual

Editor: C. ALLEN BUSH

Department of Chemistry and Biochemistry
University of Maryland Baltimore County

VOLUME 2 • 1992

 JAI PRESS INC.

Greenwich, Connecticut   *London, England*

CONTENTS

LIST OF CONTRIBUTORS

C. Allen Bush

Department of Chemistry and Biochemistry
University of Maryland Baltimore County
Baltimore, Maryland

Perseveranda Cagas

Department of Chemistry and Biochemistry
University of Maryland Baltimore County
Baltimore, Maryland

Therese M. Cotton

Department of Chemistry
Iowa State University
Ames, Iowa

Maurice R. Eftink

Department of Chemistry
University of Mississippi
University, Mississippi

Randall E. Holt

UOP
Monirex Systems
175 Oakton St.
Des Plaines, Illinois

Andrew J. Howard

Protein Engineering Department
Genex Corp.
Gaithersburg, Maryland

Jae-Ho Kim

Department of Chemistry
Iowa State University
Ames, Iowa

Thomas L. Poulos

Center for Advanced Research in
     Biotechnology of the Maryland
     Biotechnology Institute
9600 Gudelsky Dr.
Rockville, Maryland

Department of Chemistry/Biochemistry
University of Maryland
College Park, Maryland

Robert W. Woody

Department of Biochemistry
Colorado State University
Fort Collins, Colorado

INTRODUCTION TO THE SERIES:
AN EDITOR'S FOREWORD

The JAI series in Chemistry has come of age over the past several years. Each of the volumes already published contain timely chapters by leading exponents in the field who have placed their own contributions in a perspective that provides insight to their long-term research goals. Each contribution focuses on the individual author's own work as well as the studies of others that address related problems. The series is intended to provide the reader with in-depth accounts of important principles as well as insight into the nuances and subtleties of a given area of chemistry. The wide coverage of material should be of interest to graduate students, postdoctoral fellows, industrial chemists and those teaching specialized topics to graduate students. We hope that we will continue to provide you with a sense of stimulation and enjoyment of the various sub-disciplines of chemistry.

Department of Chemistry
Emory University
Atlanta, Georgia

Albert Padwa
Consulting Editor

PREFACE

This second volume continues our survey of new topics of special interest in the field of biophysical chemistry. We include an article on the technique which has been the most crucial in revolutionizing our thinking about how proteins work. The article by Howard and Poulos gives an especially compact and enlightening summary of developments in the methodology of X-ray crystallography. Optical activity, perhaps the most senior of the methods of biophysical chemistry, has assumed new importance as circular dichroism is used to follow nucleation of folding of the domains of proteins. Woody's article is a thoughtful analysis by the leading theorist of the CD of peptides. Eftink's article on the fluorescence of alcohol dehydrogenase summarizes an enormous amount of data on the system which has been used to test this method for determining the structure and dynamics of biopolymer assemblies. Although surface-enhanced Raman spectroscopy has often been associated with analytical chemistry, the article by Cotton and coworkers illustrates the power of the resonance effect in providing a highly selective probe for detailed studies of selected regions in large systems. The final article on the structure of complex carbohydrates of glycoproteins, represents a topic so new that there continues to be controversy on some of the most basic principles of confor- mation and dynamics. We have attempted to sort out some of the conflicting views on the interpretation of NMR and molecular modeling results on complex car- bohydrates.

I continue my appeal to the readers of these volumes to send me your comments regarding not only the contents of those already published, but also your ideas and suggestions for topics which I may have overlooked and which might be included in planning for the future of Advances in Biophysical Chemistry.

C. Allen Bush
Series Editor

METHODS IN MACROMOLECULAR CRYSTALLOGRAPHY

Andrew J. Howard and Thomas L. Poulos

Advances in Biophysical Chemistry, Volume 2, pages 1–36
Copyright © 1992 by JAI Press Inc.
All rights of reproduction in any form reserved.
ISBN: 1-55938-396-8

I. INTRODUCTION

The 1980s witnessed a remarkable rebirth in interest in macromolecular structure and function. This is especially true in macromolecular X-ray crystallography and 2D NMR spectroscopy. The interest in NMR spectroscopy is relatively easy to understand since it has been only in the last few years that the promise of 2D and now 3D and 4D NMR methods has been realized[1,2,3,4] with some very promising advances. These are exciting times for NMR spectroscopy as the technique is rapidly developing and proving itself a powerful method if not the only method for determining the solution structure of proteins. The reasons for a renewed interest in crystallography, however, are not so straightforward, since the promise of crystallography was realized nearly 30 years ago with the solution of the first protein structure, myoglobin.[5,6] Why, then, has there developed in recent years such intense interest in crystallography? The history of twentieth century science will probably provide an objective answer to this question, but we will seek an answer even in the absence of a long view.

If one looks at the development of macromolecular crystallography in the United States it nicely divides into three periods corresponding to the last three decades. In the 1960s several American universities established crystallography groups after the demonstrated success of the myoglobin structure. American postdoctoral fellows learned their trade at Medical Research Council Laboratories in England and returned home to establish the first American groups devoted to protein crystallography. During the 1960s, the development of diffractometers and computers made the job somewhat easier although far from routine and still a painstaking task. Nevertheless, remarkably rapid progress was made especially with respect to how enzymes work. The first Cold Spring Harbor Symposium totally devoted to macromolecular crystallography held in 1971[7] was a celebration of this first decade of structure solutions.

Interest in further growth and establishment of new groups in crystallography waned in the 1970s. The general perception was that the method was too expensive and it still took far too long to solve a structure. It was felt that it was better to devote limited university resources to the more rapidly developing areas in molecular biology. Moreover, the easy problems had been solved. That is, those proteins which were easy to obtain in large quantities and crystallize had been spoken for. Moreover, by the 1970s the perception was that we knew a good deal about globular proteins and that there may not be many more surprises. Thus, the emphasis shifted from solving the structure of anything that crystallized and diffracted well to solving the structure of proteins related to specific biological problems. The problem of purification and quantity became a much more serious obstacle.

Everything changed in the 1980s. The renaissance in crystallography and the general area of macromolecular structure and function coincided with the birth of the new biotechnology. This was not accidental. More than anything else, the breathtakingly rapid development and application of recombinant DNA technology was responsible for the renewed interest in protein structure and function. Suddenly there was a plentiful supply of precious proteins with novel, interesting, and commercially useful biological activities. Regulatory DNA binding proteins are one noteworthy example. Solving the structure of a DNA polymerase, while highly desirable when it was discovered, was unheard of 10 years ago since a monumental effort would have had to have been mounted to obtain enough protein. Now one obtains enough from less than a kilogram of a suitably engineered microorganism. In some of the better systems, it is possible to have the protein of interest account for 30–40% of the protein produced by *E. coli*. Such overexpression makes the job of purification much less formidable. It is not uncommon to have 2 or 3 step purification procedures. These advances not only have led to higher production of rare proteins but also has made available purer samples that will more readily crystallize. Now crystallographers and NMR spectroscopists have an entire new class of critically important proteins to work with.

This article will focus on the general methods required to solve and refine a macromolecular structure with special emphasis on recent methodologies and advances and on those that appear to hold the most promise in the near future. Determination of a protein structure involves four basic steps: (1) purification and crystallization; (2) data collection from parent and derivative crystals; (3) location and refinement of heavy atom positional parameters and phase calculation; (4) interpretation of electron-density maps and refinement of the structure. We will discuss each of these in turn. Macromolecular crystallography is a large and complex field, and we cannot expect to cover every aspect of it in depth (readers are invited to examine volumes 114 and 115 of *Methods in Enzymology,* referenced repeatedly below, as an expression of the state of the art as of 1985); we will concentrate on developments since 1985 and on extensions of pre–1985 methods to new contexts.

II. PURIFICATION AND CRYSTALLIZATION

The scattering of X-rays from a single molecule of a protein or nucleic acid does not produce a detectable signal, at least with the current generation of X-ray sources (see, however,[8]), so it is necessary to study samples in which the scattering is amplified through constructive interference over billions of macromolecules, all oriented and spaced in an orderly fashion. The order in the sample gives rise to scattered or diffracted rays strong enough to detect, and it is these diffracted rays that are the raw data of crystallography. The prerequisite for their creation is the ability to make the ordered arrays of macromolecules, that is, crystals. These crystals rarely arise in nature (see, however,[9]), so the production in the laboratory of these crystals is the first step in a crystallographic project.[10]

Crystallization of a macromolecule resembles crystallization of any other substance in that it involves the creation of conditions under which the lowest-energy state for a molecule is as an element of an ordered lattice of similar molecules. The crystalline state is characterized by low entropy, so the enthalpic forces favoring crystallization must be sufficient to overcome entropic tendencies toward remaining in solution.[11] All protein and nucleic acid crystals obtained to date are grown out of aqueous solution, so the equilibrium for crystallization must lie in the direction of a solid, ordered array rather than as a solution; further, the kinetics of crystal growth must favor a small number of large, fully ordered arrays rather than the production of small or disordered aggregates.

Macromolecular crystals are different from most small-molecule crystals in that the maintenance of the native structure requires the presence of water. Further, the intermolecular forces that hold the lattice together are sometimes mediated through solvent molecules. Thus the role of water (and sometimes other elements of the crystallization medium) in the kinetics and equilibrium conditions of crystallization is crucial. Crystallographers recognize both "bound" solvent molecules, which are found in well-defined positions in relation to the macromolecule, and "disordered" solvent molecules, which are present in the lattice but occupying different positions in adjoining unit cells.

In general the techniques used to crystallize macromolecules involve manipulation of a nearly-saturated or supersaturated solution. The solubility of a macromolecule depends on the concentration and identity of the cosolutes, the pH, and the temperature, all of which are subject to the control of the experimenter. If a solution can be prepared in which a macromolecule is on the verge of precipitating, some of these variables can be gradually altered so that the precipitation occurs in an ordered way rather than chaotically. The formation of the ordered precipitate is called nucleation, and the addition of further solute molecules to the nucleation site is called crystal growth. If a small number of nucleation sites can be induced by changes in the variables mentioned above, and the ordered growth of a lattice at one of these sites can be maintained long enough for a large crystal to appear, the crystallization will have succeeded. Any other result—production of too many nucleation sites, disordered growth, or premature cessation of growth—fails to produce usable crystals.

Most techniques exploit the dependence of the macromolecule's solubility on the concentration of cosolutes. For most macromolecules there is a salt concentration at which they are maximally soluble, and the most common crystallization techniques, "salting out" and "salting in", exploit the dependence of solubility on salt concentration. In a "salting out" crystallization, a protein that is maximally soluble in, say, 1 M $(NH_4)_2SO_4$ is dissolved under those conditions and then gradually diffused against a solution of 2 M $(NH_4)_2SO_4$ so that supersaturating conditions are arrived at gradually. Nucleation sites form in a few minutes to a few weeks, and crystals can grow to diffraction-quality sizes (typical linear dimensions from 0.1 mm to 1.5 mm) in one day to one year. The opposite approach of "salting in," in which a nearly-saturated solution of a protein that is more soluble at high ionic strength than at low is equilibrated against a lower-ionic strength buffer, also finds applications. Other cosolutes used in crystallizations include alcohols like ethanol and 2-methyl-2,4-pentanediol and polymers like polyethylene glycol; these agents break up clusters of water molecules and modify the water-macromolecule interactions in ways that sometimes promote crystal growth.

Until around 1980 macromolecular crystallization was seen as more of an art than a science. In a typical crystallization effort, a researcher would obtain a concentrated sample of a macromolecule and run through his or her personal favorites among the techniques on a short list of possibilities. If crystals were obtained, the project would proceed; if not, the crystallizer would try a few other techniques from the list. If one of the attempts yielded poor-quality crystals, the researcher might vary one or more of the conditions of the setup in an attempt to improve the crystals, and if all went well, diffraction-quality crystals would appear.

Two changes have improved the success rate of crystallizers. The first is a better understanding of how to approach the crystallization effort systematically, with due attention to the actual properties of the specific macromolecule. Multidimensional analyses of the factors affecting solubility and aggregation are carried out, and computer-aided techniques for stepping through the ranges of numerical variables (pH, temperature, and ionic strength) are available. Meanwhile, the "lore" of crystallization has grown richer and less anecdotal, so that the researcher can effect discontinuous changes in conditions, like the substitution of one salt for another, with some basis in past experience for the choice. To some extent, crystallization requires the hands of a good artisan and there are researchers who are still known as gifted crystallizers, but the opportunity to substitute systematic approaches for intuitive flashes and luck is more available now than it was a decade ago. Recently a 600–entry database on the crystallization of biological macromolecules has been assembled[12] and made available through the Brookhaven Protein Data Bank and in a convenient IBM PC format.

The second and perhaps more significant reason for greater success in crystallization is an improvement in the samples with which the crystallizer begins. The availability of high-performance liquid chromatography (HPLC), preparative electrophoresis, affinity chromatography, and other state-of-the-art protein purification techniques allows the protein crystallizer to begin with cleaner starting materials. HPLC has been particularly effective at increasing the number of tools in the purifier's toolbox: reverse-phase, ion-exchange, and sizing HPLC columns are all routinely available, and the simplicity and rapidity with which a researcher can pass samples through his choice of columns allow high yields with relatively small effort expended.

Molecular genetics plays a role as well. The ability to express cloned proteins in bacteria at high levels enables the researcher characterizing a protein to produce milligram quantities of soluble, uncomplexed proteins that are available in the original host only at very low yield and often complexed to carbohydrates or lipids. The cloned protein will ordinarily be a single polypeptide species, whereas the natural isolate might copurify with several almost identical isozymes. Crystallizations of heterogeneous mixtures of isozymes usually fails, so eliminating this complication improves the chances of success. Starting with a pure, homogeneous sample reduces the likelihood of an impurity disrupting a growing crystal lattice and the likelihood of an ill-understood interaction between two components of the crystallization medium that prevents nucleation or growth.

Recent developments show promise in further expanding the realm of successfully crystallized proteins. Partial or full automation of the crystallization process is underway in several laboratories, usually using commercially available robotic devices.[13] Robots of the current generation are slower than human workers and are incapable of some intricate motions. But a robot can repeat any job within its capabilities a very large number of times, so the robot can relieve the human crystallizer of much of the tedium involved in scanning wide ranges of crystallization conditions. Furthermore, robots do not get sick, so they can be used in manipulations of highly toxic proteins or proteins derived from highly toxic organisms. There is a synergy between improvements in protein purity and yield and the availability of robots for crystallization; robotic crystallizations are most effective when fairly large quantities of pure protein are available, so that many different crystallization conditions can be sampled, and those large quantities are much more available now than they used to be. An additional important factor in computer-controlled crystallization is that the computer operates as the lab notebook, thereby avoiding errors in recording the variables in a large number of experiments.

An even more exotic development is crystallization in microgravity.[14] Convection is believed to be one of the principal causes of cessation of crystal growth, and by eliminating the pull of gravity on a growing nucleation site one can reduce density-driven convection currents. Also, under some conditions sedimentation interferes with the formation of single crystals, and microgravity eliminates sedimentation. Finally, crystals grown in gravity fields tend to stick to a surface of the vessel in which they are grown, making it difficult for the researcher to separate the crystal from the surface film and from neighboring crystals. Considerable effort has therefore been focused over the last eight years on the development of apparatus for crystallizing macromolecules under zero-gravity conditions, particularly aboard the Soviet or American space shuttles. Encouraging results, including the production of diffraction-quality crystals, have been obtained from several of these efforts. Microgravity-grown porcine elastase crystals diffract better than the best available earth-grown crystals, and enable the determination of a better-quality structure.[15] Funding from NASA for the microgravity crystallization efforts and for the physico-chemical studies that underlie them has also enabled a better understanding of macromolecular crystal growth.

A few of the fundamental properties of macromolecular crystals are worthy of note. First, they are highly hydrated, and the water structure associated with the macromolecules in the lattice is essential to the crystal's stability. Second, the stability of the lattice is rapidly compromised by exposure to X-rays; thus the data collection efforts described in the next section tend to be races against time in which the experimenter attempts to measure all the relevant Bragg reflections from the crystal before radiation damage has disintegrated the lattice to the point where the crystal no longer diffracts. Finally, and most positively, crystals of macromolecules display many of the same properties that the same molecules display in solution: many enzymes remain enzymatically active in the crystalline state, and numerous indirect experiments confirm that the solution structure of a protein or nucleotide is substantially the same as the structure in the crystal. There are exceptions to this generalization, and in any case a single diffraction experiment can at best provide a "snapshot" of a molecule that is known to be flexible. But the relevance of this science stems principally from the fact that the structure determined by X-ray diffraction generally resembles at least one of the states the macromolecule adopts in solution.

III. CRYSTALLOGRAPHIC DATA COLLECTION

A. Sources

1. Conventional Sources

X-ray diffraction studies begin with mounting a crystal on an appropriate apparatus, exposing it to X-rays, and detecting the diffracted beams. There are two components to crystallographic apparatus: the X-ray source and the detection equipment. X-rays for diffraction have traditionally been obtained by training high-voltage electrons on a metal target so that the target emits X-rays at characteristic energies. The most common targets are copper and molybdenum, which emit $K\alpha$ radiation (accompanying a promoted electron's return from the L shell to the K shell) at $E = 8$ keV ($\lambda = 0.154$ nm) and $E = 17.5$ keV ($\lambda = 0.071$ nm) respectively. These energies are appropriate for diffraction studies because they are on the order of magnitude of the maximum resolution (smallest Bragg-plane spacing) obtainable from crystals. Cu $K\alpha$ radiation has been used in the vast majority of macromolecular studies until the advent of synchrotron crystallography; see below.

The most significant advance in metal-target X-ray sources in recent years has been the availability of commercial devices that contain a system for rotating the anode (target) so that it does not overheat during data collection. Typically the anode is cylindrical, and it is rotated such that a different surface of the anode is "seen" by the electron beam at different times. This enables the user to train more power on the anode without melting or pitting the surface through Joule heating. This technology was first described six decades ago,[16] but only in the last two decades has it become routinely available commercially. Rotating-anode X-ray sources tend to require more maintenance than the lower-brilliance sealed-tube sources, but this problem is alleviated considerably by the substitution in recent years of ferromagnetic seals in place of the older oil-based seals.

Metal targets emit significant fluxes away from the characteristic $K\alpha$ energy. Traditional data collection methods have required a more or less monochromatic beam, so considerable attention has been paid to methods of filtering out wavelengths other than the characteristic. The traditional technique involves placing a filter made of a metal of atomic number $(Z+1)$ between the source and the sample, where Z is the atomic number of the anode metal; this technique works because of a strong absorption of the largest component of the undesirable radiation, namely the $K\beta$ radiation. A cleaner beam may be obtained by placing a single crystal of graphite in the path of the unfiltered beam and collecting the diffracted beam emitted in the (002) direction; since different wavelengths scatter at different angles, the beam can be made almost arbitrarily monochromatic simply by selecting a range of scattering angle from the graphite. These so-called monochromator crystals are used with most area detector and diffractometer systems. The high degree of monochromaticity so obtained gives rise to data with better signal-to-noise characteristics and, due to decreased absorption of long-wavelength X-rays, longer crystal lifetime. An alternative approach, now used with most film systems and some area detectors, employs double-focusing mirror.[17] In this case bent mirrors are aligned so that the input beam is totally internally reflected from them and the resulting beam is directed at the sample. The beam is not only reasonably monochromatic; it is also intense and has a small divergence and low background. These are desirable properties, and serve to remind us that the quality of optics in a diffraction experiment is as important as how monochromatic the beam is.

2. Synchrotron Sources

A new generation of X-ray sources for diffraction studies was born in the late 1970s with the availability of X radiation from electron storage rings or synchrotrons.[18] These huge, expensive instruments are found in a small number of facilities around the world, but the special properties of the sources have aroused considerable interest. Synchrotron radiation (SR) is generated by positrons or electrons confined by magnets to move at relativistic velocities in a circle called a storage ring. Acceleration of the particles results in the emission of synchrotron radiation which is captured by beam lines tangential to the storage ring. SR has a wide spectrum of wavelengths but the intensity is orders of magnitude larger than even the largest rotating anode X-ray generators. Hence the advantages of SR are the extreme intensities and multiple wavelengths available. The wavelength available from SR is a function of the radius of the storage ring and is given by:

$$\lambda = 0.559R/E^3 \tag{1}$$

where λ is the "critical" wavelength in nm, R the radius of the storage ring in meters, and E the energy of the circulating particles in gigaelectronvolts. The large mix of wavelengths covered is between 0.05 and 0.2 nm which is precisely the region required for X-ray diffraction. The short-wavelength range (<0.1 nm) is particularly interesting, since crystal absorption is smaller so that crystal lifetimes are longer and absorption corrections are less crucial. One clear advantage, besides the high intensity, is that radiation used is "cleaner" since conventional sources rely on electrons bombarding a metal surface, usually copper. Such surfaces wear with time, leading to decreased intensity and radiation with poorer spectral characteristics.

The desired SR wavelength is obtained by first passing the beam through a single crystal of a suitable (non-protein) material that is aligned so only the diffracted X-ray at the desired wavelength is used—that is, a monochromator. Mirrors also can be used to further focus the beam. Extreme care in focusing and collimation of such an intense beam can generate extremely high quality data with higher signal to noise ratios than are available using conventional sources.

Fluxes from the existing storage ring sources at Orsay (France), Stanford (U.S.), Hamburg (Germany), Cornell (U.S.), Daresbury (UK), Brookhaven (U.S.), and the Photon Factory (Japan) are typically one to four orders of magnitude more intense than a rotating-anode generator. The new generation of storage rings represented by equipment under construction at Grenoble and Argonne will provide two additional orders of magnitude. There is a trade-off between flux and the monochromaticity of the beam, and for one instrumental arrangement at Stanford for which the wavelength spread $(\Delta\lambda/\lambda) = 10^{-5}$, the fluxes are no higher than those obtainable from a conventional source; but the compensating advantages of the tunable, ultra-monochromatic beam make that arrangement nonetheless attractive. The multi-wavelength anomalous dispersion (MAD) technique, as we will discuss below, requires this extreme tunability and monochromatization.

These intense sources increase the rate of decay of macromolecular crystals even as they increase the rate at which diffraction data can be collected. Usually the increase in the decay rate does *not* keep pace with the increase in data rate, so that the synchrotron experiment is not only speedier (i.e., the experimenter gets all of his results more rapidly) but of higher signal-to-noise quality.

3. Laue Techniques

All the experimental setups described to this point employ nearly monochromatic radiation. A radically different approach to diffraction is to deliberately use a broad spectrum of X-ray energies, and measure Bragg reflections at all energies simultaneously. White-beam experiments can be performed on conventional X-ray sources, where the non-monochromatic Bremmstrahlung ("braking") radiation from the metal target is intense enough for certain applications, but most of the recent attention has been on white-beam experiments at synchrotrons. Here the beam intensity, impressively high even at a single wavelength, becomes breathtakingly high, and the experimental arrangements are fairly straightforward. Experiments that would require days or weeks on a conventional source and minutes or hours on a monochromatic synchrotron source require milliseconds on a white-beam source. A lysozyme diffraction pattern has been recorded on a white-beam source from a single "bunch" of storage-ring electrons—10 picoseconds' worth. Crystallographers are envisioning time-resolved studies comprised of "snapshots" of protein's state as little as a few milliseconds apart. The detection and analysis aspects of these experiments will be explored below.

B. X-ray Detection

1. Traditional Techniques

Crystallographers have alternated between photographing diffracted X-rays and counting them. The earliest observation of a diffraction pattern was recorded on film,[19] but most of the earliest structure determinations were carried out using data from X-ray spectrometers. Film collection became the method of choice by the late 1930s, but the development of partially or fully automated scintillation-counter diffractometers in the 1950s and 1960s resulted in the pre-eminence of counting methods for small molecule crystallography.

For small proteins the superior statistics of a counter continue to make it an attractive choice.[20,21] Instruments are fully automatic in that a researcher can mount a sample of unknown unit cell dimensions, determine the crystal's characteristics, and measure the Bragg intensities with very little manual intervention. Full coverage of reciprocal space is facilitated by mounting the crystal on an Eulerian cradle or the more complex kappa geometry, either of which has three rotational degrees of freedom. Once the crystal is optically aligned, the instrument and its controlling computer search through reciprocal space for a few reference reflections, index them by an integerness-search technique,[22] and proceed to measure each reflection's intensity. Software is available that permits rational coverage of reciprocal space, often exploiting crystal symmetries and allowing the user to collect Friedel-related reflections (see next section) close together in time. Data processing software includes various forms of profile-fitting.[23,24] The scintillation detector used in the measurements is equipped with computer-controlled attenuators to allow accurate measurements over a wide dynamic range.

For larger macromolecular applications film has retained a large share of the market; the disadvantages of decreased accuracy and precision are more than offset by the increase in counting efficiency. In a macromolecular crystallographic experiment dozens to thousands of Bragg reflections are diffracting at any one time, so the inefficiency of a single-counter technique, for which all but one of those reflections will be lost to the experimenter, is evident. The development of screenless precession[25] and screenless oscillation photography[26] improved the precision and convenience of film techniques and made film the method of choice for crystals with unit cell dimensions >~10 nm from 1975 until recently.

In a typical screenless-oscillation experiment, the crystal is mounted on a horizontal spindle with one of its crystallographic axis aligned along the spindle. The crystal is rotated through a moderate-sized angle (typically 0.3°–1° for large proteins or viruses and 2°–4° for medium-size proteins) about this spindle while the X-rays are on and the diffracted beams are recorded onto a flat, curved, or V-shaped film cassette. The cassette usually contains a pack consisting of at least three pieces of film back to back; the first film is intended for measurement of the weakest reflections, while the rearward films permit quantitation of the brighter reflections that are saturated on the front film. After an exposure lasting 1–3600 seconds another cassette is inserted, the spindle is advanced by somewhat less than an oscillation width (to provide for some overlap) and the process is repeated. As many films are collected as necessary to measure all the unique Bragg reflections accessible in this geometry. The crystal can be reoriented to allow additional data to be collected about another axis to measure reflections that are inaccessible in the first setting. Each of the fifty or more film packs is extracted from the cassette while a subsequent oscillation is underway; the individual films are developed by hand, and the intensities of the spots on the films are measured on an optical densitometer. The background "under" each reflection is estimated from the optical density outside of, but in the vicinity of, the spot. Film–to–film scaling within a pack is accomplished by comparing unsaturated spots appearing in two or more of the films, and scaling from one pack to another relies on symmetry-related reflections or overlapped spots. Spots whose scanning-angle profiles or rocking widths span two or more of the oscillation ranges are usually measured on the pack in which they are strongest, and a refinable estimate of the "degree of partiality", that is, the portion of the total profile actually contained within that primary film, is obtained in order to compute the total intensity of the spot.[27] As with diffractometry, profile-fitting techniques have enabled researchers to extract weak signals out of the high and poorly-determined backgrounds characteristic of macromolecular data.[28,29] For applications such as synchrotron data collection, film data collection remains the most common technique, but the cumbersomeness of the development and film-scanning steps are contributing to the demise of this method.

2. Electronic Position-Sensitive Detectors

Numerous researchers have recognized that the ideal measuring device for macromolecular diffraction data would combine the area detection capabilities of film with the precision and accuracy of a counter. This combination has to some degree been achieved in the development of electronic area detectors, and for many crystallographers these have become the preferred instruments for data collection.

An electronic area detector is any instrument that is capable of converting photon events occurring over an extended area into electronic signals. As such the category includes both photon counters, in which each photon event gives rise to a signal that is recorded as a single count at a specified location on the face of the instrument; and analog detectors, for which the arrival of a photon on an X-ray sensitive phosphor gives rise to light photons, which are then "read out" in some fashion. Prototypes of both types of instrument were developed in the 1970s, and the first structures were solved with them by mid-decade. Commercial instruments became available in the early 1980s, and over 100 laboratories worldwide now have crystallographic area detectors. The two classes of detectors are described in further detail below.

The first photon detectors successfully used for crystallography were flat multiwire proportional counters. These devices were first developed in the 1950s and 1960s for high-energy physics applications, and Charpak[30] described their potential application to detection of soft X-rays like those used in crystallography. The instrument is a chamber containing xenon gas, with two planes of wires, one horizontal and one vertical, fixed within it. Each photon arriving through the gas-tight window of the detector gives rise to ionizations of the xenon in specified locations, and the ionization is recorded as an electronic event on the horizontal and vertical wires near that location. These electronic events are recorded with a delay-line readout, so that the circuitry recognizes each photon event as a discrete count at a single (X,Y) location. In the crystallographic application, a stationary crystal is exposed to X-rays for an interval while counts are being recorded; at the end of the interval the data in a two-dimensional histogram of (X,Y) positions is read into a computer or stored on a disc, and then the crystal is moved to a different position. The data read into the computer is treated as a single two-dimensional electronic "image" of the diffraction pattern and analyzed in software. Instruments of this type, developed at the University of California, San Diego[31] and at the University of Virginia[32] have been used in solving many protein structures; the San Diego design is now available commercially (San Diego Multiwire Systems, Inc.)

These instruments are fairly large (24 cm on a side) and compensate for their relatively low spatial resolution (typical pixel size ~ 1 mm × 2 mm) by covering large amounts of solid angle. These detectors can be used in multi-detector arrays, further increasing the amount of solid angle covered. Because the detectors are flat, they can be used over a wide range of crystal-to-detector distances. This affords an advantage in data collection over wide ranges of unit cells: the detector is set close to the crystal if the unit cell is small, and farther away if the cell is larger. The diffracted beams increase in both size and separation as one moves the detector back from the crystal, but the increase in separation is faster than the increase in size, so closely-spaced spots are better resolved farther back. For a 7 nm unit cell the detector is typically placed about 40 cm away; for a 22 nm unit cell it is placed over 1 m away.

A related design is the spherical drift chamber. Here the ionizations derived from photons that pass through the window are directed along radial paths through a gas-filled drift region and are detected at points within a flat detection region. As with the flat detectors, the photon events give rise to electronic pulses on planes of wires in the detection regions, and these are read in by a fast but position-sensitive fashion. A detector of this type has been developed at MIT,[33] and another is currently in use on the synchrotron at LURE in Orsay, France;[34] like the San Diego and Virginia designs, these instruments have active areas of about 25 × 25 cm. The radial paths in the drift region require a defined distance from the crystal to the detection region, so the crystal-to-detector distance is fixed; therefore, it is not possible to collect data from large unit cells (>15 nm). For some applications, however, the geometrical precision of this instrument makes it more suitable than a flat detector.

A smaller photon counter with some of the characteristics of both of the above classes was developed commercially in the early 1980s[35] and is available commercially (Siemens Analytical X-ray Instruments, Madison, WI). This instrument, originally termed an "imaging proportional counter," uses a curved detection region like a spherical drift chamber, but the drift region is compact enough that it can be used over a wide range of crystal–to–detector distances. The ionizable gas in this instrument is pressurized to 4 atm so that the probability of interactions in the first 1–2 mm of the gas is very high; as with the other photon counters the count causes electronic events on a pair of wire grids. A current-division scheme is used to read out the electronic events. This detector is run at a low gain so that the ion cloud induced by each photon event is small. Therefore the instrument has a higher spatial resolution than the larger detectors (point-spread FWHM ~ 1 pixel ~ 0.2 mm).

All of these proportional counter designs suffer from dead-time losses when the count rate is high. If the detection electronics is still processing a count when another arrives, the identification of the (X, Y) position for one or both of the counts may fail. The commercial detectors rely on readout schemes for which these "coincidence losses" become serious around 50 kHz; revamping the readouts could increase this value to perhaps 1 MHz, but fluxes from the existing synchrotron sources surpass this value, and the sources currently under construction are altogether out of range. So the analog devices, described below, are the detectors of choice on storage rings.

In the 1970s the most extensive development of two-dimensional electronic analog detectors occurred in the laboratory of Dr. Uli Arndt at Cambridge.[36] The Cambridge group developed small (8 × 8 cm) detectors with a X-ray sensitive phosphor coupled through a photomultiplier to a silicon-intensified target (SIT) television camera. Most of the components of the system are available commercially, contributing to the flexibility and low cost of the system. The Cambridge system allows for readout in the form of discrete, complete electronic "frames," or through sampling only selected three-dimensional (X, Y, and scanning angle) blocks of data. It is now available commercially (Enraf-Nonius, Delft, Netherlands). Similar developments at Brandeis[37] and Princeton Universities[38] have not seen extensive application in single-crystal diffraction studies.

A new generation of analog detector designs relies on charge- coupled devices (CCDs) rather than SIT tubes for light detection. Thus crystallographic detector development has followed patterns already followed in astrophysics, where photography gave way first to SIT-tube designs and then to CCD-based designs; the difference is that the visible light is not the raw data, but rather an outcome of the conversion in the phosphor of X-rays to visible light. For the most part the technology developed for SIT-tube systems carries over to the CCD's, which appear to offer greater thermal stability and improved geometrical uniformity than the earlier technology.[39]

A radically different area detection technique has received considerable interest in the last half-decade.[40,41] Detectors known as "image plate" systems consist of a fairly large Gd_2O_2S phosphor coupled to a sensor that reacts to light by generating excited states in "color centers" consisting of europium atoms embedded in a passive matrix. The "image" of excited states can be read out fluorescently by scanning the plate with an appropriately tuned gas laser in a separate readout step, and the fluorescence output can be recorded digitally. Thus an image plate is not an electronic detector at all, since the imaging and digitization is performed offline without recourse to electronic readout, but it has an end product that is similar to that produced by electronic detectors—a two-dimensional "image" of the diffraction pattern recorded while the image plate is being exposed. The image plate behaves rather like film in that it requires a somewhat cumbersome readout step; the differences are that the noise levels are much lower and the readout consumes no chemicals.

Each of these detector designs offers some advantages and some disadvantages. The large, flat multiwire chambers display a uniform quantum response and nearly linear geometric response, but they are somewhat cumbersome to manipulate and their relatively low spatial resolution makes them difficult to use with large unit cells. The spherical drift chambers display no parallax and their response is fairly uniform, but they cannot be moved at all and are difficult to use with large unit cells. The high-pressure, low-gain small detectors have good spatial resolution and are easy to manipulate, so they are well-suited to large unit cells; but they have highly nonlinear geometric characteristics and their quantum response is somewhat less uniform than with the flat chambers. The SIT-tube television detectors can be run at arbitrarily high input flux, have fairly good geometrical uniformity and are well-interfaced with the crystal-mounting hardware, but the high dark current and the sensitivity to changes in temperature and magnetic fields limit their accuracy. The image plate detectors have good spatial resolution relative to the detector size, and they are not count-rate limited; but the need for a separate readout step introduces significant and time-consuming complications in the data processing. In general, all of these detector designs can be and have been used to solve numerous structures, and the decision of any particular laboratory to use or purchase one design or another is often based less on its underlying characteristics and more on external issues like the software that is bundled with it or the cost and after-sale service.

3. Area Detector Software

The volume of raw data generated with an area detector system—typically on the order of 100 megabytes per day of data collection—requires that the data be processed by computer. The successful determination of structures with area detectors has required the development of software to turn this huge volume of raw data into accurate, precise intensity measurements. Until the mid-1980s software for area detectors was fairly tightly associated with specific detector designs, and the software priorities have been chosen in response to the needs of those detectors. Thus a package particularly tuned to the characteristics of the San Diego-style detectors was developed at San Diego,[42] a SIT-tube package was developed at Cambridge,[43] and so on. More recently a package designed for use with several different kinds of area detectors (including, in principle, film) has been developed by J. Pflugrath and others with partial funding from the EEC;[44,45] a package with fairly general application has been developed by L. Weissman, and the *XENGEN* package currently used in most laboratories containing the Siemens Area Detector[46] is now being generalized to handle data from other types of detector.[47]

Area detector software, whether tied to a particular instrument or not, must be responsible for the following tasks:[48]

- Determination of the orientation of the crystal, that is, the directions of the unit cell axes of the crystal in a laboratory reference frame.
- Refinement of the crystal's unit cell parameters and the orientational information.
- Determination of the intensity and variance of each reflection appearing within the data that have been collected.
- Correction for systematic error in the intensity measurements—errors due to crystal decay, absorption, and instrumental peculiarities.

In addition, specific packages offer additional capabilities, including utilities to facilitate motion of the crystal about rotational degrees of freedom, video displays of the electronic images generated during data collection, feedback on diffraction-spot predictions and diffraction quality during data acquisition, outlier rejection code, and statistical analyses. In several cases the computer responsible for controlling data acquisition is not the computer performing the data processing; this was unusual in the days of monolithic, standalone computers, but in the distributed computing environment of today it is far from surprising.

The first of the major tasks—determination of the crystal's orientation—may be accomplished with varying degrees of automation. On most of the early systems the orientation was determined by exposing the crystal to X-rays, viewing the resulting diffraction patterns, and rotating the crystal about the axes on which it was mounted to bring it into a position where a recognizable pattern could be obtained. More recent software packages have included auto-indexing routines, which determine the orientation by exploiting the integerness of the sampling of reciprocal space.

Refinement involves minimization of an identifiable residual, usually involving differences between observed and predicted centroids in detector coordinates (X,Y) and in scanning angle. This refinement may employ an ordinary conjugate-gradient least-squares procedure or a simplex algorithm. Data input to this refinement may come from specially-constructed "refinement" data runs or from a sample from the data to be used in the integration itself.

Once the crystal orientation and unit cell are known to the user and the software, the determination of intensities can begin. Most early software packages employed a straightforward summation technique, in which a Bragg reflection's position on the detector face and the scanning angle were determined beforehand, and the background-subtracted counts in the images close to that position were summed. Various instrumental and experimental corrections were applied to this sum and the result was taken to be the integrated intensity of the reflection. Recent software has supplemented the summation technique with some form of profile-fitting,[49] in which the intensity is determined by reference to a predicted three-dimensional model profile. In either summation or profile-fitting techniques, the crucial issues for data quality include precise tracking of the crystal orientation, typically through "on-the-fly" re-refinement of the orientational parameters, and the accuracy and robustness of the background estimate. Intensities of macromolecular diffraction spots are often very low, so trustworthy background estimates are particularly crucial to data quality.

The high speed with which area detector data can be collected obviates some of the messier problems of crystallographic scaling, such as the indeterminacy engendered by merging data from large numbers of crystals, but correction for systematic error remains a crucial problem. Macromolecular crystals are often fairly large (linear dimensions ~0.5 mm), and their tendency to form undesirable shapes (plates or needles) means that corrections for absorption are large in magnitude and rapidly varying with scan angle through the crystal. Most macromolecular crystallographic data collection is carried out with the crystal stuck to the inside wall of a glass or quartz capillary, with an irregular puddle of crystal mother liquor attached to it. Therefore the absorption from components of input and output beam paths is considerable. Even with an area detector it is customary to collect data on a crystal until X-ray decay has reduced intensities down to

50–80% of the initial values, and the decay is usually fastest for the highest-resolution data. Suitable models in software for correcting these sources of systematic error are critical to the success of a crystallographic effort. Various approaches[50,51] have been proposed and coded, and the "best" approach or approaches are still under investigation. We believe the ideal approach will combine an appropriate scaling technique, in which the parameters of the scaling functions are determined by minimizing the scaled differences between the intensities of symmetry-related reflections, with an external experiment that characterizes the absorption profile of the crystal and the beam path. The external experiment can provide a correction for the gross absorption profile, and the symmetry-based scaling can deal with the fine details. Several mathematical forms of the scaling parameterization will work about equally well, but an expansion in spherical harmonics is stable numerically and has some physical reality, so it is a suitable choice.

C. Laue Experiments

In a Laue experiment, the crystal is illuminated with multiple wavelengths at one time. This results in many Bragg planes satisfying the requirements for diffraction at a single setting of the crystal. Hence, there is no need for motion of the crystal as with conventional monochromatic sources. Since a large region of reciprocal space is covered at a single setting, nearly a complete data set can be obtained with one exposure. Until the mid-1980s it was thought that harmonic overlap would render multiple wavelength white-beam crystallography impractical. In any polychromatic X-ray diffraction experiment, certain Bragg reflections will appear precisely overlapped upon one another, as a reflection (h,k,l) that satisfies the Bragg condition at wavelength λ overlaps $(2h,2k,2l)$ at wavelength $\lambda/2$, and (nh, nk, nl) at wavelength λ overlaps (mh, mk, ml) at wavelength $(n\lambda/m)$. Recent theoretical and practical studies have shown that for a typical macromolecular crystal the percentage of the data affected by these harmonic overlaps is less than 17%, so the majority of the Bragg reflections can have their intensities measured by more or less conventional techniques.[52] Therefore a resurgence of interest in these white-beam or Laue techniques has arisen in recent years. As mentioned above, the enormously high fluxes attainable with white beam techniques, coupled with the ability to collect a complete or nearly complete data set from a single photograph, allows crystallographers to design time-resolved crystallographic experiments, in which one measures a series of "snapshots" of the course of a chemical reaction, each "shot" a few milliseconds long and spaced perhaps a few tenths of a second apart. A pilot experiment of this type was carried out on glycogen phosphorylase B at Daresbury; the complex of the enzyme with a substrate, maltoheptaose, was sampled at three time points, each requiring three seconds of data collection.[3] In the long run the most serious difficulty experimenters will encounter in this type of experiment is crystal decay: the flux incident on the crystal for each exposure is so high that decay will be appreciable even on the millisecond time scale, and by the tenth 50 msec exposure of the crystal it will not be diffracting as well as on the first.

Many of the challenges remaining to the application of the white-beam technique to routine crystallographic experiments are analytical. The crystal's absorption varies rapidly as a function of energy; these variations must be carefully corrected for in data processing or else "scaled away" by calculation of fractional difference maps, in which the amplitudes used in the Fourier transform are:

$$\Delta F_{hkl} = [(F_{hkl}(\text{reac}, \lambda) - F_{hkl}(\text{nat}, \lambda))/F_{hkl}(\text{nat}, \lambda)] * F_{hkl}(\text{nat}, \lambda_0)$$

where $F_{hkl}(\text{nat}, \lambda_0)$ is the ordinary native structure amplitude at the monochromatic wavelength, $F_{hkl}(\text{nat}, \lambda)$ is the native structure amplitude observed in the Laue experiment at its appropriate wavelength, and $F_{hkl}(\text{reac}, \lambda)$ is the Laue amplitude for the reactive intermediate of interest, that is, the amplitude measured at the specific time point under consideration. These difference amplitudes, coupled with the model phases for the enzyme, yield interpretable difference maps that can image the current state of the protein-ligand complex.

Other analytical problems include harmonic and spatial overlaps, indexing, and strategy. Harmonic overlaps must be recognized in software and either eliminated from consideration or deconvoluted; experimental deconvolution methods are available for film data collection but not for image plate collection, and the possibility of entirely analytical deconvolutions exists. Unlike monochromatic experiments, in which spatial overlap of spots is unlikely except along very long unit cell axes, the admixture of wavelengths found in a Laue film results in close approaches of nonharmonically overlapped reflections toward one another. Like harmonic overlaps, these must be recognized in software and either eliminated from integration or deconvoluted.[54] The high density of spots on a film makes indexing, that is, identification of the *hkl* indices of each spot, a non-trivial problem, and elaborate graphics-aided techniques of assuring correct indexings have been derived.[55] In cases where the crystal symmetry is low enough that one film is insufficient for collection of a complete data set, one needs a way of assuring that no more films be collected than are necessary; this involves a sophisticated understanding of reciprocal space, and given the "high stakes" involved in these expensive experiments, the need for software-aided strategies is evident. In spite of all these problems, the prospects for solving structures and following reaction courses over millisecond time scales using Laue diffraction represents the most exciting advance in macromolecular crystallography in the last half-decade, and the full promise of this method is likely to be realized before the end of the century.

IV. METHODS OF PHASE DETERMINATION

A. Multiple Isomorphous Replacement (MIR)

A fundamental equation in crystallography used to calculate electron density maps is the following Fourier summation where the triple summation is over all *hkl* values.

$$\rho(x,y,z) = 1/V \sum \sum \sum F_{hkl} \exp[-2\pi i(hx + kl + lz)] \tag{2}$$

Here ρ = electron density, V = volume of unit cell, $\mathbf{F}_{hkl}$ is the structure factor, and *hkl* are the Miller indices. Each reflection or spot on an X-ray photograph corresponds to a specific *hkl* and structure factor, $\mathbf{F}$. The quantity measured in an X-ray diffraction experiment is the intensity, I, of each *hkl* reflection. I in turn is proportional to the square of the structure factor, $\mathbf{F}^2$. However, $\mathbf{F}$ is a complex number consisting of both an amplitude ($\mathbf{F}$) and phase (α), $\mathbf{F}_{hkl} = \mathbf{F}_{hkl} \exp i\alpha_{hkl}$. The product of a complex number with its conjugate

$$[\mathbf{F}_{hkl} \exp(i\alpha_{hkl})][\mathbf{F}_{hkl} \exp(-i\alpha_{hkl})] = \mathbf{F}_{hkl}^2 \tag{3}$$

results in the loss of phase information. The experimental observable is thus a phaseless intensity. Herein lies the fundamental problem in crystallography— determination of the phases.

The method of multiple isomorphous replacement (MIR) pioneered by Perutz and Kendrew over 30 years ago is still the only method routinely used for initial estimates of phases for a new structure. In this method, a large atom (mercury, platinum, etc.) is diffused into the protein crystal. If the heavy atom binds such that the crystalline lattice and protein structure are minimally perturbed, the measured intensities obtained from the derivative will differ slightly from the parent. This is because the structure factor of the heavy atom derivative, $\mathbf{F}_{PH}$ is the sum of the parent and heavy atom structure factors, $\mathbf{F}_{PH} = \mathbf{F}_P + \mathbf{F}_H$. Because this is a summation of vectors, it is often easier to visualize the phase problem using vector diagrams or Argand diagrams an example of which is shown in Figure 1.

We measure the amplitude of $\mathbf{F}_P$ and $\mathbf{F}_{PH}$ which allows an estimate for the amplitude of $\mathbf{F}_H$. What is required is the location of the heavy atom(s) within the unit cell. This usually is accomplished using Patterson techniques. The Patterson function is the Fourier transform using the square of the amplitudes as coefficients and shows peaks at the ends of vectors connecting atomic centers. In a protein such a Patterson map would be hopelessly complex because there would be thousands of peaks of about equal intensity. Heavy atoms, however, dominate such a map. Therefore, if coefficients of the form $(\mathbf{F}_{PH} - \mathbf{F}_P)^2$ are used, interatomic vectors involving heavy atoms will dominate, allowing the location of the heavy atoms in the unit cell. Once we know the location, α_H is known. In a perfect world, the above triangle closes enabling a good estimate of all values including α_P. The one

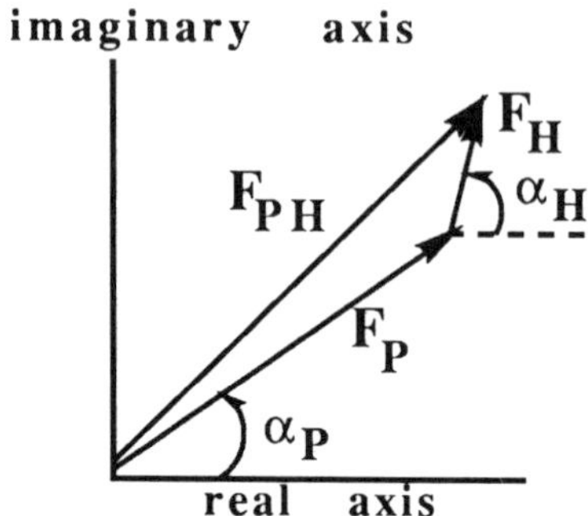

Figure 1. An argand vector diagram showing the relationship between the parent structure factors (F_P), the heavy atom structure factor (F_H), and the sum of the two. α_H and α_P represent the phase of the parent and heavy atom, respectively. since these are complex numbers, there is both an imaginary and a real axis.

ambiguity remaining is the sign of the phase. This ambiguity can be resolved with a second derivative. How this works in practice is illustrated with the vector construction shown in Figure 2. Diagrams like this are usually called Harker diagrams, after David Harker who first introduced the method in 1956.[56]

First, a circle is drawn with a radius F_P corresponding to the parent structure factor. We know the length of this vector but not its direction (phase). Second, the vector $-F_H$ (heavy atom structure factor) is added to F_P and a second circle of radius F_{PH} is drawn with its origin at the end of $-F_H$. This gives two circles that intersect at two places corresponding to the two possible phase angles. F_P and F_{PH} are obtained from the measured intensity data and the amplitude of F_H is estimated

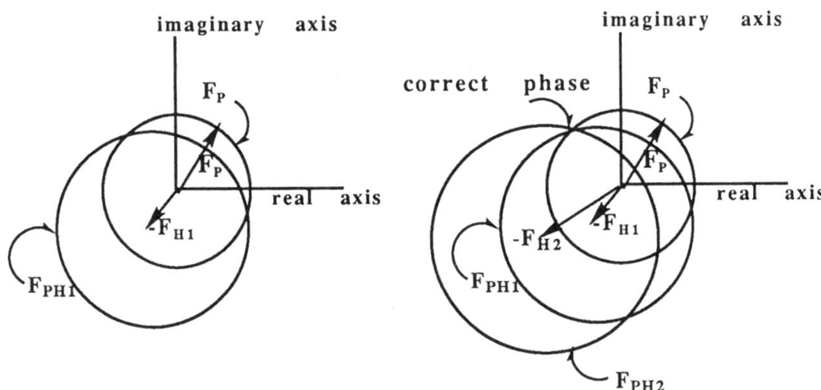

Figure 2. Harker diagrams illustrating how the phase problem is solved using isomorphous heavy atom substitution.

from the difference between $\mathbf{F}_{PH}$ and $\mathbf{F}_P$ while the position of the heavy atom and hence its phase usually is estimated using difference Patterson techniques. The positional parameters (coordinates, occupancy, and temperature factor) of the heavy atom(s) are refined by various methods usually by minimizing the difference between calculated and observed structure factors. The ambiguity in determining phases with a single derivative can be resolved with a second heavy atom derivative. Here the same circle construction is used for the second derivative ($\mathbf{F}_{PH_2}$). The correct phase is located where the three circles intersect. Alternatively, the following analytic expression can be derived:

$$\mathbf{F}_{PH}^2 = \mathbf{F}_P^2 + \mathbf{F}_H^2 + 2\mathbf{F}_P\mathbf{F}_P\cos(\alpha_P - \alpha_H)$$

$$(4)$$

We measure $\mathbf{F}_{PH}$ and $\mathbf{F}_P$ and calculate $\mathbf{F}_H$ and α_H which then allows for an estimate of α_P. While modern computer programs use similar analytical expressions to estimate phases, the method of constructing circles to determine MIR phases was carried out manually for 2000 reflections in obtaining the first low resolution electron density map of myoglobin.[57]

In the real world phase triangles don't close and the circles don't intersect at one place. Therefore, Blow and Crick[58] suggested casting the phase as a probability distribution of the form:

$$P_{\mathrm{iso}}(\alpha) = \exp(-\,|\,\mathbf{F}_{PH} - \mathbf{F}_{PH}(\mathrm{calc})\,|\,E_{iso}^2)$$

$$(5)$$

where the quantity $|\,\mathbf{F}_{PH} - \mathbf{F}_{PH}(\mathrm{calc})\,|$ is the lack of closure error referring to how poorly the phase triangle closes and E represents cumulative errors. This has been the primary method used for estimating MIR phases for over 30 years. The errors, and hence problems, arise primarily from lack of isomorphism, too little or too much heavy atom binding, errors in intensity measurements, and scaling of derivative and parent data sets. These problems usually manifest themselves in errors in determining $\mathbf{F}_H$. Next to crystallization, the estimate of initial phases is the most uncertain and frustrating aspect of a protein structure solution.

While there have been some sophisticated improvements in data quality and computer methods of refining heavy atom positions and calculating phases, the basic MIR technique has remained unchanged since it was first used with myoglobin. Trial-and-error seems to remain the method of choice. However, recombinant DNA methods offer a way for designing specific heavy atom sites. The most successful heavy atom derivatives often involve covalent modification of cysteine sulfhydryl groups with mercurials. Therefore, if a protein lacks SH groups, site directed mutagenesis can be used to replace a serine with cysteine. Alternatively, if a protein contains many SH groups leading to too many heavy atom sites, the problem of positioning heavy atom sites can be made easier by removing SH groups.

B. Molecular Replacement

If the new structure to be solved closely resembles that of a protein whose X-ray structure is known, molecular replacement methods can be-used and the MIR method avoided. In this case only a parent data set is required. The general idea is to position the known structure into the unit cell of the unknown structure. Then structure factors (amplitudes and phases) can be calculated from the coordinates of the known structure and conventional refinement methods used to solve the structure of the unknown. Refinement methods adjust the atomic positional parameters to minimize the R factor which is:

$$R = \sum | \mathbf{F}_o - \mathbf{F}_c | / \sum \mathbf{F}_o \tag{6}$$

where $\mathbf{F}_o$ = observed amplitudes obtained from the unknown, and $\mathbf{F}_c$ = calculated amplitudes obtained initially from the known atomic coordinates and in the latter stages of refinement the unknown coordinates. The first step is to find the proper orientation of the known structure. This is accomplished with the rotation function defined as follows:[9]

$$R(C) = \int P_1(\mathbf{x}) P_M(C\mathbf{x}) dV \tag{7}$$

where P_1 is the Patterson of the unknown crystal and P_M is the Patterson of the known molecule. $R(C)$ is sampled at various orientations, C (usually represented by three Eulerian angles). Where peaks occur in the rotation function represent possible correct orientations of the known structure into the unknown unit cell. Part of the Patterson function consists of self vectors (i.e., interatomic vectors between atoms within a molecule) and depends only on the structure. The rotation function, R, will reach a maximum when a self vector of the known structure matches that of the unknown. Obviously, the more similar the two structures the less ambiguous the result.

The next step is to translate the known structure, which is now (we hope) in the correct orientation, to correctly position the known in the unknown unit cell. Where the rotation function depends on intramolecular vectors in Patterson space, the so-called translation function depends on intermolecular vectors between molecules related by the space group symmetry of the unknown. As the known molecule is stepped through the unit cell of the unknown, intermolecular Patterson vectors will reach a maximum when the known is correctly translated. Alternatively, "brute force" methods can be used if a large computer is available. Here the molecule is stepped through the unit cell and structure factor calculations carried out at each position. The calculated structure factor, $\mathbf{F}_c$, of the known is then compared with the observed structure factors for the unknown, $\mathbf{F}_{obs}$. Where $\mathbf{F}_{obs} - \mathbf{F}_c$ is a minimum represents a possible solution. The general usefulness of molecular replacement methods has been significantly enhanced by the availability of a "user friendly" software package known as MERLOT.[60]

The main limitation of molecular replacement is that the known and unknown structures must be homologous such as two globin structures or two c-type cytochromes. Molecular replacement cannot be used for a new structure whose fold is significantly different from any of the known protein structures or where common elements of secondary structure are oriented significantly differently. Nevertheless, molecular replacement has proven a powerful method for obtaining structural information on related proteins without the need for arduous heavy atom derivative methods. All one needs is a single good parent data set.

C. Multiple Wavelength Anomalous Dispersion (MAD)

Another method for estimating phases takes advantage of the fact that heavy atoms not only diffract X-rays but also absorb X-rays. When the frequency of an incoming X-ray is close to an absorption frequency of a heavy atom, the X-ray will experience a phase shift and become attenuated. This effect is called anomalous scattering or dispersion. Because of this effect, the atomic scattering factor (f_o) must be corrected by real (f') and imaginary ($i\Delta f''$) terms. The real term usually is a negative number loosely correlated with attenuation of the X-ray beam while the imaginary term represents the phase shift.

$$f = f_o + f' + i\Delta f'' \tag{8}$$

How this affects structure factors is illustrated with the vector diagram shown in Figure 3.

$\mathbf{F}_L$ is the structure factor without anomalous scattering, $\mathbf{F}_H$ is the contributions of the heavy atoms to the structure factor, and $\mathbf{F}_{hkl}$ and $\mathbf{F}_{-(hkl)}$ are the Friedel related structure factors. Note that f' is subtracted from $\mathbf{F}_H$ because in most cases f' is a negative number. Also note that f'' is normal to $\mathbf{F}_H$ which illustrates the phase shift associated with anomalous scattering. Without a significant anomalous effect,

$$\mathbf{F}_{hkl} = \mathbf{F}_{-(hkl)} \tag{9}$$

This equality (Friedel's Law), however, does not hold when there is a significant anomalous scattering center(s). The difference in Friedel-related reflections is usually very small, but if these differences can be accurately measured useful phase information can be extracted.

The situation is strictly analogous to the MIR method. The same set of circles and diagrams used for the MIR method can be constructed. By accurately measuring the differences between Friedel-related reflections, the position and hence phase of the anomalous scattering center(s) can be estimated. Anomalous differences are used most frequently to break the phase ambiguity associated with a single derivative. Therefore, it is sometimes possible to find a unique solution to the phase problem with one heavy atom derivative that has a large anomalous signal.

There is now increasing optimism that anomalous dispersion can be used for a direct estimate of phases without the need for heavy atom derivatives if the parent

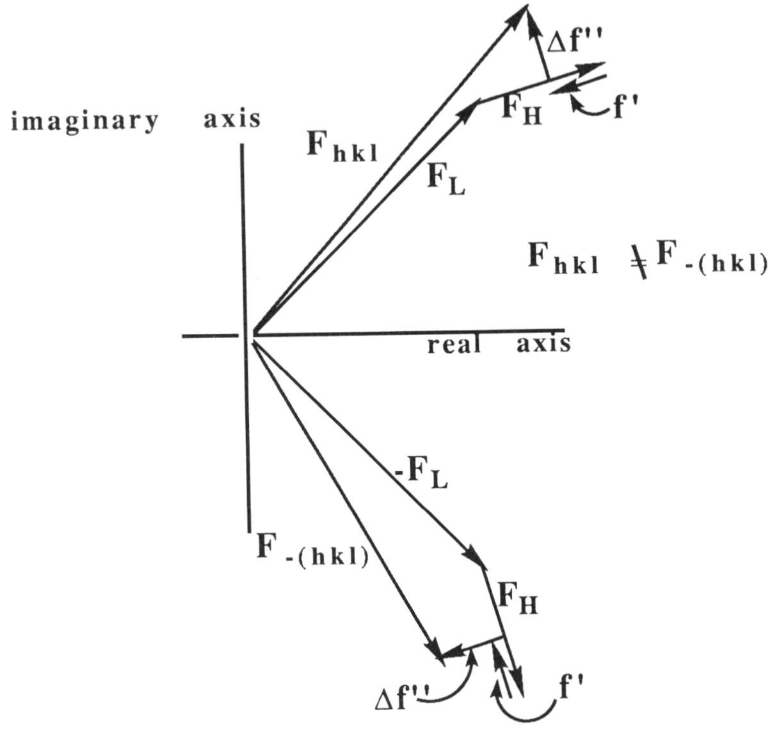

Figure 3. A vector diagram illustrating the effect of anomalous dispersion on structure factors.

protein contains a heavy atom such as iron. For proteins without naturally occurring heavy atoms, a single derivative can be prepared and phases determined directly from the single derivative alone without the need for a parent data set or other derivatives. There are obvious advantages to this approach. It totally avoids the problem of nonisomorphism, scaling of derivative to parent data sets, and the often tortuous search for several heavy atom derivatives. The key to this method is to collect data sets at several different X-ray wavelengths. The effect of multiple wavelength data sets is exactly the same as multiple derivatives. The easiest way to illustrate how this works is to describe an actual experiment using MAD to estimate phases for a protein. Harada et al.[61] used the synchrotron source at the Photon Factory in Japan to obtain phases for cytochrome c' using the iron anomalous signal. The vector diagram in Figure 4 illustrates how phases were derived.

Here $OA = \mathbf{F}_L$, the structure factor of the light atoms and $AB = \mathbf{F}_H$, the structure factor of the heavy atom, in this case iron. BC represents f' (the real component of

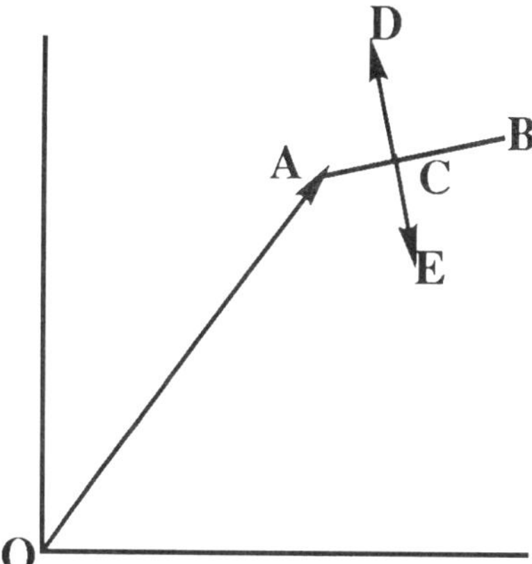

Figure 4. Vector diagram outlining how anomalous dispersion and multiple wavelengths can be used to solve the phase problem.

the anomalous scatterer), and $CD = CE = \Delta f''$, the imaginary component. Data were collected at three wavelengths, λ_1, λ_2, and λ_3. The wavelengths selected satisfied the following set of criteria. For λ_1 and λ_3, $\Delta f''$ (the imaginary component of anomalous scattering) were small and can be ignored. The real component of the anomalous scatterer, f', also is small and negligible for λ_1. For λ_3 anomalous scattering is maximum since both f_0 and f' are large. Hence, $\mathbf{F}_{+1} = \mathbf{F}_{-1} = OB$, $\mathbf{F}_{+3} = \mathbf{F}_{-3} = OC$, $\mathbf{F}_{+2} = OD$ and $\mathbf{F}_{-2} = OE$. Therefore, with four different sets of structure factors ($\mathbf{F}_1$, $\mathbf{F}_{+2}$, $\mathbf{F}_{-2}$ and $\mathbf{F}_3$) one has the same amount of information required for MIR phase determination. Sets of circles are drawn for each structure factor and where they intersect is the correct solution to the phase problem.

To fully exploit the method requires the collection of intensity data at different X-ray frequencies in the frequency range where the anomalous scattering effect is maximum. Although the potential for using multiple wavelengths and anomalous scattering to obtain initial phasing information has been known for some time,[62,63] most conventional X-ray sources do not operate in appropriate frequency ranges. Renewed interest in MAD is due to the availability of tunable synchrotron radiation sources and two dimensional detector systems for rapid data collection (see previous section).

It is necessary to obtain experimentally derived values for $\Delta f''$ and f' as a function of wavelength since these are influenced by the environment of the anomalous scattering center and hence will be different for each protein. This is accomplished by first measuring the fluorescence spectrum from a single crystal around the absorption edge of the anomalous scatterer. The fluorescence data are obtained with a detector positioned normal to the X-ray beam used for the diffraction experiment. Therefore, the same crystalline sample and experimental set up can be used for both diffraction and fluorescence experiments. This also enables an experimental determination of which wavelengths should be used for data collection. $\Delta f''$ can be determined from the fluorescence data since it is directly related to the atomic absorption coefficient. However, f' cannot be directly measured but can be calculated since f' is related to $\Delta f''$ by the Kramers-Kronig transformation. As in the cytochrome c′ example above, one wavelength is selected where $\Delta f''$ is a maximum and another where $\Delta f''$ is small but f' is maximally negative. Care also must be taken when operating near the absorption edge of the heavy atom since the effect of crystal shape becomes an important variable. X-ray absorption will be maximum when the beam passes through the thickest part of the crystal and obviously becomes a problem for asymmetrically shaped crystals.

Although the MAD method of phasing has found limited use, several studies have clearly shown the promise of this method. The iron atom of heme proteins[64] and the copper atom of azurin[65] have been used for MAD phasing and computation of interpretable electron density maps. However, phasing on naturally occurring metal centers will be useful for relatively small proteins since the anomalous signal for iron or copper is quite small compared to metals typically used for MIR phasing.

MAD may become more generally useful for proteins if at least one derivative can be prepared. In this case, the electron density of the derivative is calculated and its structure solved. Then derivative calculated phases can be used to solve the parent structure using straightforward refinement techniques. The feasibility of this approach has been demonstrated for a $TbCl_3$ derivative of parvalbumin.[66]

Overall, MAD offers distinct advantages over traditional MIR methods for obtaining primary phase information. All data are obtained from the same parent or derivative crystals so the problem of nonisomorphism and scaling is minimized as much as is possible. Coupled with recombinant DNA techniques where heavy atom sites can be "custom designed", MAD should prove to be a powerful technique. The primary limitation on the wider use of MAD is accessibility to the required instrumentation which is available at only a few installations. In the long run, however, problems with using such multiuser resources may far outweigh the often time consuming search for several heavy atom derivatives.

D.   Direct Methods

The Nobel Prize in Chemistry for 1985 was awarded to J. Karle and H. Hauptmann for the development (by them, D. Sayre, M. Woolfson, and others) of direct methods in crystallography.[67,68] These techniques have turned the process of solving small-molecule crystal structures into a largely automatic procedure, in which structure amplitudes are determined by diffractometric measurements and phases are determined by reference to relationships ("triplet relations", "quartet relations", and so on) among groups of strong reflections. The technique is called a "direct method" because both the amplitudes and the phases are derived from a single experiment. These concepts are embodied in several widely-distributed software packages and allow for successful structure determinations in the majority of cases where the molecular weight of the sample is less than 1000. The application of these techniques to macromolecules has met with limited success so far, in part because the number of reflections that need to be phased is so large and also because the data are inherently weak and noisy. A more fundamental limitation in these methods is that in most formulations they depend on atomicity, that is, on having diffraction measurements at sufficiently high resolution (small Bragg plane spacings) that individual atoms can be distinguished from one another. Very few proteins diffract measurably beyond 0.15 nm, and the ability to distinguish individual atoms is borderline at that resolution.

Recent work has suggested applications for direct methods in macromolecular crystallography, however. Several researchers have integrated direct methods techniques with single isomorphous replacement, that is, they have developed formalisms in which triplet relations are used in conjunction with data from a single isomorphous derivative to generate phase estimates.[69,70] Others have focused on applications of the direct-methods formalism to the problem of anomalous-dispersion phasing. Some of the mathematics in the multiple wavelength technique described above owes to these direct-methods procedures.[71]

Several researchers are suggesting ab initio phasing techniques that employ a maximum-entropy formalism. Harrison[72] has developed such an *ab initio* phasing technique, and has successfully solved several small molecule structures and ultimately the structure of an 8-kilodalton DNA oligomer.[73] Other approaches include examination of electron density histograms, within which only those that correspond to physically meaningful distributions of electron density are accepted.[74] Each of these "direct" phasing techniques remains in the experimental stage, and for the most part protein crystallographers turn to these techniques only when "tried and true" approaches—multiple isomorphous replacement and molecular replacement—fail.

E. Density Modification

Any of the methods described above can give rise to phase estimates of sufficient quality to enable the crystallographer to begin to fit an experimental electron density map to a structural model. However, the phase errors at this stage are large, and procedures for improving the phase estimates that do not rely on an accurate tracing of the chain are valuable. These techniques generally fall into the heading of "density modification" techniques, because they rely on modifications of the electron density values based on rudimentary chemical knowledge.

The simplest form of density modification is solvent flattening; here the modeler notes that, outside an envelope defined by the outer reaches of the macromolecule, the electron density should be a constant equal to the density of the bulk solvent in the crystal. Thus, if a spherical protein molecule of radius 3 nm sits near the center of a crystallographic asymmetric unit 8 nm on a side, then within the 8 nm box, but outside the 3 nm sphere, the electron density should be a constant ρ determinable from the crystallization conditions. If we modify our error-prone electron density map (obtained from Eq. (2)) by replacing the experimental electron density with this value ρ at all external points and perform the reverse Fourier transform, improved phase estimates may arise.

Similarly, in a noisy map there are often locations at which the experimental electron density is negative; we can simply truncate those density values to zero (or a small positive value), invert the Fourier transform, and look for improvements in phases. Typically the map that results from the next "round" of Fourier map generation displays clearer connectivity than the un-modified map, usually (but not always) simplifying the processs of map-fitting.

These density modification techniques have been embodied in a popular software package and have been used in the early stages of many macromolecular structure determinations.[75] An adaptation of the density-modification concept to incorporate a maximum-entropy formalism[76] has proved useful in other cases.

V. MODEL FITTING AND REFINEMENT

Two closely intertwined tasks remain to the crystallographer once he or she has obtained an initial set of experimental phases. The first is to examine the electron density map obtained from the Fourier transform of the complex structure factors F_{hkl} and assign a set of atomic coordinates consistent with that electron density map. This process is known as "model fitting" or "map fitting," depending on whether one wishes to emphasize the source of the information or the desired result, and it can be carried out with varying degrees of automation. The second is to improve the phase estimates and the resulting electron density map by adjustments in the modeled electronic coordinates. This process is known as "structure refinement," and it is generally carried out by a calculation in software of the adjustments necessary to minimize the differences between the observed structure amplitudes F_{hkl} (obs) and the amplitudes F_{hkl} (calc) obtained from a Fourier inversion of the coordinate model. The two processes are complementary, as symbolized in Figure 5.

In spite of their theoretical complementarity, as a practical matter the processes of model fitting and refinement are performed separately. The algorithms required to approach these problems in software are very different, and even the kinds of computing equipment required is somewhat different; model-fitting is most easily accomplished with a graphics workstation capable of displaying the molecular model in pseudo-three-dimensional form, whereas refinement requires little more than appropriate algorithms and parameters, plus raw processor power.

A. Model Building

Until the early 1970s the most effective technique for fitting an atomic model

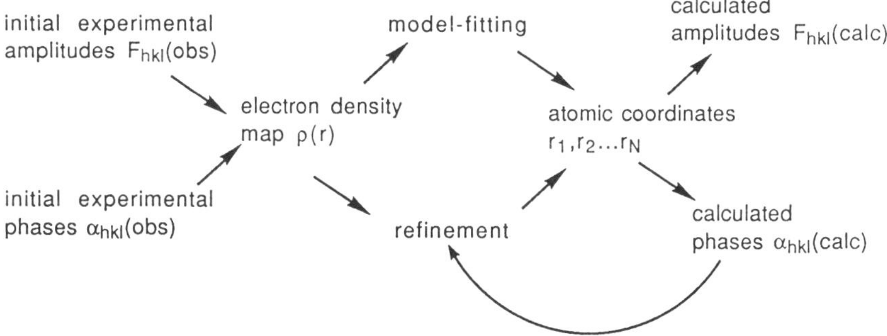

Figure 5. General outline of a structure determination and refinement.

to an electron density map involved building a brass mechanical model of the molecule at a scale of about 25 cm/nm. The builder matched the coordinates of the atoms in the model to the map with a device known as a "Richards Box" containing computer-generated transparencies of the map and a half-silvered mirror.[77] The method was accurate, but it was slow and cumbersome, and the finished product was difficult to examine. A large protein model could be more than 2 m on a side, and the innermost atoms would be difficult to see through the forest of rods (representing covalent bonds) and junctions (representing atoms). A faster and more flexible model-building technique was needed, and the solution was computer graphics.

Two-dimensional graphical display devices have been available for three decades, and by the late 1960s efforts were underway to provide an approximation to three-dimensional displays through depth cueing, stereoscopic images, and color. Research at Princeton, UC San Francisco, IBM, Washington University, and the University of North Carolina focused on developing hardware and software tools for building atomic models by computer graphics. With pseudo-three-dimensional graphics devices a user can display a chosen portion of a contoured electron density map and use his chemical intuition and understanding to build an atomic model that fits the map. Commercial hardware is now available for display and manipulation of molecular models (Evans & Sutherland, Salt Lake City, UT; Silicon Graphics, Mountain View, CA; Stardent, Santa Clara, CA). Early graphical display devices were slaved to a mainline computer or minicomputer so that most of the arithmetic calculations required for the model-building would be performed away from the graphics processor, but current offerings are in fact full-function workstation computers as well as display devices.

Graphics hardware used in molecular modeling is equally applicable to other pseudo-three-dimensional contexts, such as airplane design and meteorology, but the software required for molecular modeling is specialized. Crystallographers and theorists began developing software for model-building and model-fitting in the early 1970s, and the development has continued through the present. The preeminent software for this purpose is FRODO[78] and its successors, O[79] and CHAIN. These packages have elaborate user interfaces allowing the user to make many choices as he builds; the convenience of the software is important to its utility because the builder may be spending hundreds of hours at the graphics device, moving bonds, rerouting atomic connectivity, and adding or subtracting solvent molecules from the model. The model-building software includes the ability to do a form of structural regularization during the building[80] or even an on-the-fly least-squares refinement.

The "classical" approach to model-building outlined above involves a good deal of user input. Researchers at UNC, Uppsala, and elsewhere have devised methods to automate the model-building to varying degrees, thereby reducing the need for time-consuming and error-prone human intervention. An algorithm known as "ridgeline tracing" builds atomic connectivity by attending to gradients in electron density; it has been successful in several instances. Other researchers have applied artificial intelligence techniques to structural model-fitting with some success.[81] Recent successes in predicting the secondary structures of proteins via neural networks[82] could be used as a starting-point for model-fitting. A more promising approach capitalizes on the availability of information from already-determined macromolecular structures. In this approach the user accesses a library of structural "fragments", typically two to twenty amino acids or nucleotides in length, and he inserts a fragment that fits well into the density under scrutiny.[83,84] The fragment library may be built up from the Brookhaven Protein Data Bank[85] or from a private database.

B. Structure Refinement

Small molecule crystallographers routinely subject their structures to full-matrix least-squares refinement in order to improve the quality of the structure. In this technique, a residual:

$$R = \sum_{hkl} | \mathbf{F}_{hkl}(\text{obs}) - \mathbf{F}_{hkl}(\text{calc}) | \Big/ \sum_{hkl} \mathbf{F}_{hkl}(\text{obs}) \qquad (10)$$

is minimized with respect to the coordinates and thermal parameters of the atoms in the structure. The minimization is based on a model in which the electron density around an atom i ($i = 1, \ldots, N$) at r_i is approximated as a Gaussian ellipsoid.

$$\rho(r) = \rho_0 \exp(- \| B_i \cdot (r - r_i) \|)^2) \qquad (11)$$

where B_i is a symmetric tensor characterizing the atom's "thermal ellipsoid." The calculated structure factors are then determined according to the ordinary inverse Fourier transform:

$$\mathbf{F}_{hkl}(\text{calc}) = \int \sum_i \rho(r) \exp(-2\pi i h \cdot r) \, d_3 r \qquad (12)$$

where the integration is carried out over the volume of the crystal's unit cell. From this equation a normal matrix is constructed; its elements are the derivatives of the structure factors with respect to the atomic coordinates r_i and the thermal tensors B_i. In some cases the thermal tensor is assumed to be a multiple b_i of the identity matrix so that Equation (11) reduces to:

$$\rho(r) = \rho_0 \exp(-b_i | (r - r_i) |)^2) \tag{13}$$

This type of refinement is reasonably stable and fast on a molecule with up to fifty atoms, but it becomes much trickier on a macromolecule. A stable least-squares refinement requires that the number of raw data F_{hkl} be several times the number of parameters being refined; with a tensorial thermal model this number will be $N * (3 + 6) = 9N$, and even with scalar thermal parameters it will be 4N. Macromolecular crystals rarely diffract to better than a minimum plane spacing (maximum resolution) of 0.15 nm, and for a typical unit cell this resolution does not provide enough data for a full-matrix refinement of 9N parameters; it is borderline for 4N. Even when the overdeterminacy (ratio of data to parameters) is adequate, full-matrix techniques are very slow on all but the fastest computers. Recent successes with full-matrix least-squares approaches on small proteins[86] suggest that they will have a place in the crystallographer's toolkit, but other techniques are important with macromolecules.

Some of the earliest macromolecular refinements employed a "real-space" technique,[87] in which a structure is refined by minimizing:

$$\int \left(\rho(r) - \rho_m(r) \right) d_3 r \tag{14}$$

(where ρ is the density obtained from the Fourier transform of the observed structure factors and ρ_m is the electron density obtained from the model. This technique met with some success in the 1970s, particularly with relatively good diffractors.[88]

Major progress in refinement was made when a restrained conjugate gradient least-squares technique was promulgated and coded in the late 1970s.[89,90] In this method, the user's stereochemical understanding (ideal values for bond lengths, bond angles, and so on) is introduced into the refinement as additional observations (analogous to the observed structure amplitudes themselves), so that the problem of under-determination alluded to above is combatted not by reducing the number of parameters but by increasing the number of appropriately weighted observations. Restrained least-squares refinements are generally carried out on a subset of the list of "free" parameters as well: hydrogen atoms, which scatter X-rays weakly and are therefore difficult to image, are generally omitted from the refinement unless $D_{min} < 0.14$ nm, and for low-resolution ($D_{min} > 0.3$ nm) data a further reduction in number of parameters is accomplished by replacing individual atomic thermal parameters b_i with an overall thermal parameter b. Restrained refinement became the norm in the 1980s, and until the appearance of dynamics-aided refinement techniques (see below) it was rare for a structure to be refined any other way.

Conjugate-gradient refinement has yielded many successes in the last fifteen years, and it remains the method of choice in many laboratories. But it is slow in the sense that it generally cannot move an atom a long distance from its current position; the refinement-residual "well" in which an atom sits is typically only about 0.1 nm wide, so large atomic motions are precluded. The user must intervene every five to ten cycles of refinement, display a calculated electron-density map, and manually reposition portions of the structure that the refinement program cannot "un-stick" from incorrect positions.

The next major advance in refinement technology arrived with the recognition that crystallographic information could be incorporated into molecular dynamics simulations. Molecular dynamics is a technique of analyzing structures in which the coordinates and velocities of atoms are allowed to vary according to Newton's laws of motion. Thus if the initial coordinates of a set of atoms $r_i(t = 0)$ $(i = 1, \ldots , N)$ are determined from a crystallographic experiment or by other means and if their initial velocities $v_i(t = 0) = dr_i/dt$ are set by assigning random directions to them and magnitudes according to:

$$v_i^2(t = 0) \propto T \tag{15}$$

where T is a nominal temperature, then the structure will change[91] according to:

$$a_i(0) = d^2r_i/dt^2 = \left(\sum_k F_{ki} \right)/m_i \tag{16}$$

where F_{ki} is the force of type k acting on atom i with mass m_i. If the sampling in time is small enough then we can approximate the displacement and velocity at time Δt as:

$$v_i(\Delta t) = v_i(0) + \Delta t^* a_i(0) \tag{17}$$

$$r_i(\Delta t) = r_i(0) + \Delta t^* v_i(0) \tag{18}$$

For subsequent time points $(2\Delta t, 3\Delta t, \ldots)$ the process just described is repeated. For typical dynamics simulations Δt is set between 0.1 and 2 fs, and the simulation is allowed to "unfold" for between 0.5 and 1000 ps. In conventional dynamics the forces F_{ki} are derived from harmonic potentials associated with covalent bond lengths and angles and with electrostatic interactions. In crystallographic dynamics these conventional forces are supplemented with a pseudo-force restraining the calculated structure factors to be similar in magnitude to the observed structure amplitudes, so that the pseudo-potential energy function to be minimized becomes:

$$E = E_{phys} + \left(\sum_{hkl} \{F_{hkl}(obs) - F_{hkl}(calc)\}^2 \right)/\sigma_x^2 \tag{19}$$

where E_{phys} is the includes all the normal chemical interactions involved in the dynamics problem (electrostatics, van der Waals forces, and so on), and σ_x is a weight factor. Minimization of the crystallographic residual becomes, then, a part of a larger process whereby the dynamic system comes to equilibrium. Crystallographic dynamics has been embodied in two powerful software packages, X-PLOR[92] and GROMOS[93], and has been used in refining the structures of several proteins and oligonucleotides. The principal advantages offered by the dynamics approach to refinement as compared with conventional restrained least-squares techniques are that it requires less manual manipulation of the model and that it is capable of making larger adjustments in atomic positions. In the best of circumstances a refinement that would have required scores of cycles of conventional refinement, interspersed with perhaps a dozen manual refittings of the map can be replaced with ten runs of dynamics refinement with only three manual refitting sessions. The dynamics approaches are considerably newer, however, and more attention must be paid to proper establishment of the run-time parameters (such as σ_x, above) and the initial conditions of the simulation than is true with conventional refinement. Also, dynamics refinement is computationally expensive; to run a 100-picosecond simulation on a protein with molecular weight of 25 kilodaltons requires about 2×10^{12} computer cycles, or about two days on a dedicated 12-MIPS computer.

Even with modern, dynamics-aided refinement the input of the crystallographer is critical. None of the efforts to automate either the refinement step or the model-building step has successfully obviated the experience and skill of the chemist. Conversely, some structures have been published with significant errors in them, principally because of errors in interpretation by the crystallographer, often in the connectivity of otherwise well-traced chains near the surface of the macromolecule. These errors should not be viewed as evidence of incompetence of the scientist or of the invalidity of the method; rather, they are illustrations of the need for care and critical thinking about the data under scrutiny. The potential for misinterpretation of data becomes much greater if the quality of the raw data (both the directly determined structure amplitudes and the indirectly derived phases) is low. So improvements in the quality of the raw data through stronger X-ray sources, better detector technology, and more clever and robust data processing software will reduce the likelihood of human error during model-fitting and refinement, and increase the confidence held by biochemists in the results of crystallographic structure determinations.

REFERENCES

1. Wuthrich, K. *NMR of Proteins*; Wiley, N.Y., 1986.
2. Fesik, S. W.; Zuiderweg, E. R. P. *J. Magn. Reson.*, **1988**, *78*, 588.
3. Marion, D.; Kay, L. E.; Sparks, S. W.; Torchia, D. A.; Bax, A. *J. Am. Chem. Soc.* **1989**, *111*, 1515.

4. Kay, L. E.; Clore, G. M.; Bax, A.; Gronenborn, A. M. *Science* **1990**, *249*, 411.

5. Kendrew, J. C.; Bodd, G.; Dintzis, H. M.; Parrish, R. G.; Wycoff, H. *Nature* **1958**, *181*, 662.

6. Kendrew, J. C.; Dickerson, R. E.; Strandberg, B. E.; Hart, R. G.; Davies, D. R.; Phillips, D. C.; Shore, V. C. *Nature* **1960**, *185*, 422.

7. *Cold Spring Harbor Symposia on Quantitative Biology.* **1971**, *36*, passim.

8. Sayre, D.; Fankuchen Lecture, American Crystallographic Association Annual Meeting, Seattle, 1989.

9. Eisenberg, D. S.; Baker, T. S.; Suh, S. W.; Smith, W. W. In *Photosynthetic Carbon Assimilation* (1978), Siegelman, H. W.; Hind, G. Eds.; Plenum: London, 1978.

10. McPherson, A. *Preparation and Analysis of Protein Crystals*; Wiley: New York, 1982.

11. Arakawa, T. and Timasheff, S. N. *Methods in Enzymology* **1985**, *114*, 49.

12. Gilliland, G. L. *J. Crystal Growth* **1988**, *90*, 51.

13. Ward, K. B.; Perozzo, M. A.; Zuk, W. M. *Proc. Second Intl. Conf. on Protein Crytal Growth*; Giegé, R. et al., Eds. North Holland: Amsterdam, 1988; p. 35.

14. DeLucas, L. J.; Smith, C. D.; Smith, H. W.; Vijay-Kumar, S.; Sendahi, S. E.; Ealick, S. E.; Carter, D. C.; Snyder, R. S.; Weber, P. C.; Salemme, F. R.; Ohlendorf, D. H.; Einspahr, H. M.; Clancy, L. L.; Navia, M. A.; McKeever, B. M.; Nagabhushan, T. L.; Nelson, G.; McPherson, A.; Koszelak, S.; Taylor, G.; Stammers, D.; Powell, K.; Darby, G.; Bugg, C. E. *Science* **1989**, *246*, 651.

15. Navia, M. A.; McKeever, B. M. 48th Ann. Pittsburgh Diffraction Conf., Huntsville, AL., 1990, paper 2.7.

16. Muller, J. *Proc. Roy. Soc.* **1925**, *125*, 507, In *Protein Crystallography;* Blundell; Johnson, Eds.; Academic Press: New York, 1976.

17. Phillips, W. C.; Rayment, I. *Methods in Enzymology,* **1985**, *114*, 316.

18. Fourme, R.; Kahn, R. *Methods in Enzymology* **1985**, *114*, 281.

19. Friedrich, W.; Knipping, P.; Von Laue, M. *Proc. Bavarian Acad. Sci.* **1912**, 303, Reprinted in *Naturwissenschaften* **1952**, *39*, 367, and quoted in *Protein Crystallography*; Blundell; Johnson, Eds.; Academic Press: New York; 1978.

20. Wyckoff, H. W. *Methods in Enzymology* **1985**, *114*, 330.

21. Fletterick, R. J.; Sygusch, J. *Methods in Enzymology* **1985**, *114*, 386.

22. Sparks, R. A. In *Crystallographic Computing Techniques*; Ahmed, F. R., Ed.; Munksgaard: Copenhagen, 1976; p. 452.

23. Diamond, R. *Acta Crystallogr.* **1969**, *A25*, 43.

24. Oatley, S.; French, S. *Acta Crystallogr.* **1982**, *A38*, 537.

25. Xuong, Ng.-h.; Freer, S. *Acta Crystallogr.* **1971**, *B27*, 2380.

26. Arndt, U. W.; Champness, J. N.; Phizackerley, R. P.; Wonacott, A. J. (**1971**), *J. Appl. Cryst.* **1973**, *6*, 457.

27. Winkler, F. K.; Schutt, C. W.; Harrison, S. C. *Acta Crystallogr.* **1976**, *A35*, 901.

28. Rossmann, M. G. *J. Appl. Crystallogr.* **1979**, *12*, 225.

29. Filman, D. J.; Hogle, J. M. *Science* **1985**, *229*, 1358.

30. Charpak, G.; Bouclier, R.; Bressani, T.; Favier, J.; Zupancic, C. *Nucl. Instrum. & Methods* **1968**, *62*, 262.

31. Cork, C.; Fehr, D.; Hamlin, R.; Vernon, W.; Xuong, Ng.-h.; Perez-Mendex, V. *J. Appl. Crystallogr.* **1974**, *7*, 319.

32. Sobottka, S. E.; Cornick, G. G.; Kretzinger, R. H.; Rains, R. G.; Stephens, W. A.; Weissman, L. J. *Nucl. Instrum. & Methods* **1984**, *220*, 575.

33. Bolon, C.; Deutsch, M.; Lanza, R.; Quigley, G.; Rich, A. *IEEE Trans. Nucl. Sci.* **1979**, *26*, 146.

34. Kahn, R.; Forme, R.; Caudron, B.; Bosshard, R.; Benoit, R.; Bouclier, R.; Charpak, G.; Santiard, J. C.; Sauli, F. *Nucl. Instrum. Methods* **1980**, *172*, 337.

35. Durbin, R. M.; Burns, R.; Moulai, J.; Metcalf, P.; Freymann, D.; Blum, M.; Anderson, J. E.; Harrison, S. C.; Wiley, D.C. *Science* **1986**, *232*, 1127.

36. Arndt, U. W. *Nucl. Instrum. Methods* **1978**, *152*, 307.

37. Kalata, K. *Methods in Enzymology* **1985**, *114*, 486.

38. Milch, J. R.; Gruner, S. M.; Reynolds, G. T. *Nucl. Instrum. Methods* **1982**, *201*, 43.

39. Strauss, M. G.; Naday, I.; Sherman, I. S.; Kraimer, M. R.; Westbrook, E. M.; Zaluzec, N. J. *Nucl. Instrum. Methods* **1988**, *A266*, 563.

40. Hendrickson, W. A.; Ward, K. B. "Imaging Plate Detectors for Synchrotron Radiation," Howard Hughes Medical Institute Scientific Conference Center, Coconut Grove, FL., 1987.

41. Ogata, C. M.; Hendrickson, W. A.; Gao, X.; Patel, D. J. *Abstr. M2*, Amer. Crystallogr. Assoc. Annual Meeting, Seattle; 1989.

42. Xuong, Ng.-h.; Freer, S. T.; Hamlin, R.; Nielsen, C.; Vernon, W. *Acta Crystallogr.* **1978**, *A34*, 289.

43. Thomas, D. J. *Nucl. Instrum. Methods* **1982**, *201*, 27.

44. Messerschmidt, A.; Pflugrath, J. W. *J. Appl. Crystallogr.* **1987**, *20*, 306.

45. G. Bricogne, Ed. *Proceedings of the EEC Cooperative Workship on Postion-Sensitive Detector Software, Phases I & II and Phase III.* Orsay, France, 1986; CNRS/LURE.

46. Howard, A. J.; Gilliland, G. L.; Finzel, B. C.; Poulos, T. L.; Ohlendorf, D. H.; Salemme, F. R. *J. Appl. Crystallogr.* **1987**, *20*, 383.

47. Howard, A. J., Final Project Reports to NSF Small Business Innovation Research Phase I Project on grant ISI-8960411.

48. Howard, A. J.; Nielsen, C.; Xuong, Ng.h. *Methods in Enzymology* **1985**, *114*, 452.

49. Kabsch, W. *J. Appl. Crystallogr.* **1988**, *21*, 916.

50. Weissman, L. In: Computational Crystallography, **1988**, Sayre, D. Ed., Clarendon: Oxford, 1982; p. 56.

51. Takusagawa, F. *J. Appl. Crystallogr.* **1987**, *20*, 243.

52. Cruickshank, D. W. J.; Helliwell, J. R.; Moffat, K. *Acta Crystallogr.* **1987**, *A43*, 656.

53. Hadju, J.; Acharya, K. R.; Stuard, D. I.; McLaughlin, P. J.; Bardord, D.; Oilonomakos, N. G.; Klein, H.; Johnson, L. N. *EMBO J.* **1987**, *6*, 539.

54. Shrive, A. N.; Clifton, I. J.; Hadju, J.; Greenhough, T. J. *J. Appl. Crystallogr.* **1990**, *23*, 169.

55. Helliwell, J. R.; Habash, J.; Cruickshank, D.W. J.; Harding, M. M.; Greenhough, T. J.; Campbell, J. W.; Clifton, I. J.; Elder, M.; Machin, P.A.; Papiz, M. Z.; Zurek, S. *J. Appl. Crystallogr.* **1989**, *22*, 483.

56. Harker, D. *Acta Crystallogr.* **1956**, *9*, 1.

57. Bodo, G.; Dintzis, H. M.; Kendrew, J. C.; Wyckoff, H. W. *Proc. R. Soc. London Ser A* **1959**, *253*, 70.

58. Blow, D. M.; Crick, F. H. C. *Acta Crystallogr.* **1959**, *12*, 794.

59. Rossmann, M. G.; Blow, D. M. *Acta Crystallogr.* **1967**, *15*, 26.

60. Fitzgerald, P. A. M. *J. Appl. Crystallogr.* **1988**, *21*, 273.

61. Harada, S.; Masanori, Y.; Murakawa, K.; Kasai, N. *J. Appl. Crystallogr.* **1986**, *19*, 448.

62. Okaya, Y.; Pepinsky, R. *Phys. Rev.* **1956**, *103*, 1645.

63. Herzberg, A.; Lau, H. S. M. *Acta Cryst.* **1965**, *22*, 24.

64. Hendrickson, W. A.; Smith, J. L.; Phizackerley, R. P.; Merritt, E. A. *Proteins* **1988**, *4*, 77.

65. Korzun, Z. R. *J. Molec. Biol.* **1987**, *196*, 413.

66. Khan, R.; Fourme, R.; Bosshard, R.; Chiadmi, M.; Risler, J. L.; Didberg, O.; Wery, J. P. *FEBS Lett.* **1985**, *179*, 133.

67. Karle, J. *Science* **1986**, *232*, 837.

68. Hauptman, H. *Science* **1986**, *233*, 178.

69. Hauptman, H. *Acta Crystallogr.* **1982**, *A38*, 289.

70. Furey, W.; Chandrasekhar, K.; Dyda, F.; Sax, M. *Acta Crystallogr.* **1990**, *A46*, 560.

71. Hendrickson, W. A.; Smith, J. L.; Phizackerley, R. P.; Merritt, E. A. *Proteins* **1988**, *4*, 77.

72. Harrison, R. W. *Acta Crystallogr.* **1989**, *A45*, 4.

73. Miller, M.; Harrison, R.; Appella, E.; Wlodawer, A.; Sussmann, J. L. *Nature* **1988**, *334*, 85.

74. Lunin, V. Y.; Urzhuntsev, A. G.; Skovoroda, T. P. *Acta Crystallogr.* **1990**, *A46*, 540.

75. Wang, B. C. *Methods in Enzymology* **1985**, *115*, 90.

76. Prince, E.; Sjolin, L.; Alenljung, R. *Acta Crystallogr.* **1988**, *A44*, 216.

77. Richards, F. M. *Methods in Enzymology* **1985**, *115*, 157.

78. Jones, T. A. *Methods in Enzymology* **1986**, *115*, 157.

79. Jones, T. A.; Zou, J.-Y.; Cowan, S. W.; Kjeldgaard, M. Amer. Crystallogr. Assoc. Annual Meeting: New Orleans, 1990; Abstr. Q06.

80. Hermans, J.; McQueen, J. E. *Acta Crystallogr.* **1974**, *A30*, 730.

81. Greer, J. *Methods in Enzymology* **1985**, *115*, 206.

82. Kneller, D. G.; Cohen, F. E.; Langridge, R. *J. Mol. Biol.* **1990**, *214*, 171.

83. Jones, T. A.; Thirup, S. *EMBO J.* **1980**, *5*, 819.

84. Finzel, B. C. Amer. Crystallogr. Assoc. Annual Meeting: Abstr. Q05.

85. Bernstein, F. C.; Koetzle, T. G.; Williams, G. J. B.; Meyer, E. F. Jr.; Brice, M. D.; Rogers, J. R.; Kennard, O.; Shimanouchi, T.; Tasumi, M. *J. Mol. Biol.* **1977**, *112*, 535.

86. Teeter, M. M.; Zhou, R.-S. Amer. Crystallogr. Assoc. Annual Meeting: Seattle, 1989, Abstr. HB4.

87. Diamond, R. *Acta Crystallogr.* **1971**, *A27*, 436.

88. Deisenhofer, J.; Steigemann, W. *Acta Crystallogr.* **1975**, *B31*, 238.

89. Hendrickson, W. A.; Konnert, J. H. In *Computing in Crystallography,* Indian Acad. Sci.; Bangalore, 1980; p. 13.01.

90. Hendrickson, W. A. *Methods in Enzymology* **1985**, *115*, 252.

91. Brünger, A. T. *J. Mol. Biol.* **1988**, *203*, 252.

92. Brünger, A. T. *J. Mol. Biol.* **1988**, *203*, 803.

93. Gros, P.; van Gunsteren,. W. F.; Hol. W. G. J. *Science* **1990**, *249*, 1149.

CIRCULAR DICHROISM AND CONFORMATION OF UNORDERED POLYPEPTIDES

Robert W. Woody

Advances in Biophysical Chemistry, Volume 2, pages 37–79
Copyright © 1992 by JAI Press Inc.
All rights of reproduction in any form reserved.
ISBN: 1-55938-396-8

I. INTRODUCTION

Among the conformations which polypeptide chains can assume, those having a well-defined structure such as the α-helix and β-sheet have CD spectra which are well-characterized and understood, at least in broad outline.[1,2] By contrast, the unordered conformation has a CD spectrum which has provoked controversy and resisted theoretical interpretation.[2] Renewed interest in the unordered conformation of polypeptide chains has been stimulated by investigations of small peptides and fragments of proteins which can, to a significant extent, assume an ordered conformation in aqueous solution.[3,4] The unordered conformation represents a "baseline" for estimating the extent of structure formation. In the study of protein folding from an unordered state, the question of whether the starting point is fully unfolded or retains some residual ordered regions is of great interest.[5]

Recent CD[6] and vibrational circular dichroism (VCD)[7–9] studies, combined with the current interest in protein folding,[3,4,10] make a review of the CD of unordered peptides timely. In this review, I will describe the conflicting interpretations of the CD of unordered peptides, discuss some recent developments and try to present a current view of the conformation of these systems.

II. CIRCULAR DICHROISM OF UNORDERED POLYPEPTIDES

A. Early Experimental Results

Early ORD and CD studies utilized poly(Glu) and poly(Lys) in neutral aqueous solution as models for unordered polypeptides. Holzwarth and Doty[11] reported two characteristic features of the CD of these ionized polypeptides: (1) a strong negative band near 200 nm ($[\theta]_{max} \sim -40,000$ deg cm^2/dmole) (Figure 1), comparable in amplitude to the α-helix CD bands; (2) a significant wavelength shift between the CD maximum (ca. 200 nm) and the absorption maximum (ca. 191 nm). These features are surprising in a putative random coil and led to suspicions that the unordered form of poly(Glu) and poly(Lys) may deviate significantly from a random coil.

Schellman and Schellman[12] quoted W. Kauzmann as remarking in 1951 that the optical rotation of the "freely coiling" polypeptide chain implies a restricted conformation. Kauzmann estimated that the optical rotation at the NaD line is about an order of magnitude larger than would be expected for unrestricted bond rotations. As Schellman and Schellman noted, this observation agrees well with the conformational analysis of Ramachandran and coworkers[13,14] who, on the basis of hard-sphere potentials, concluded that only 8% of conformational space is fully allowed to peptide residues with β carbons, and only 23% is partially allowed. Schellman and Schellman[12] went on to suggest that ". . . the high levorotation of

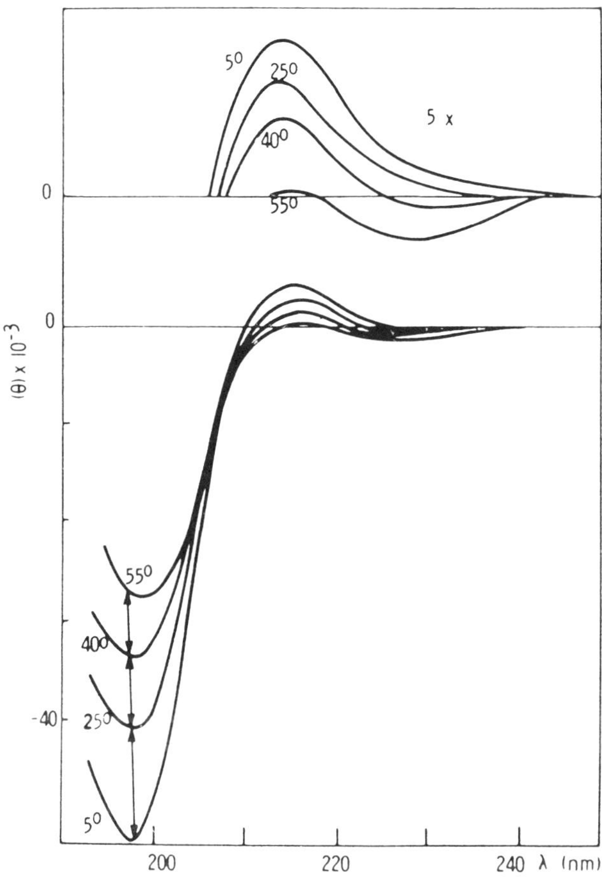

Figure 1. The CD spectrum of poly(Glu) in water as a function of temperature. (Reproduced from Tiffany and Krimm, *Biopolymers* **1972**, *11*, 2309–2316 with permission. Copyright © 1972 John Wiley and Sons, Inc.)

the random polypeptide chain might arise from a considerable probability for the orientation angles associated with the collagen fold, which has a very high levorotation . . . " Beychok,[15] citing the intensity of the 200 nm CD band and the displacement of its maximum from that of the absorption band, stated ". . . it is far from certain that these molecules [unordered polypeptides] are totally free of ordering or occasional periodicity of unknown kind."

Numerous studies of synthetic polypeptides, denatured proteins, and natural and synthetic peptides have verified that the unordered conformation invariably gives rise to a strong negative band near 200 nm, although the intensity varies widely. It should be noted, however, that a strong negative band near 200 nm does not always

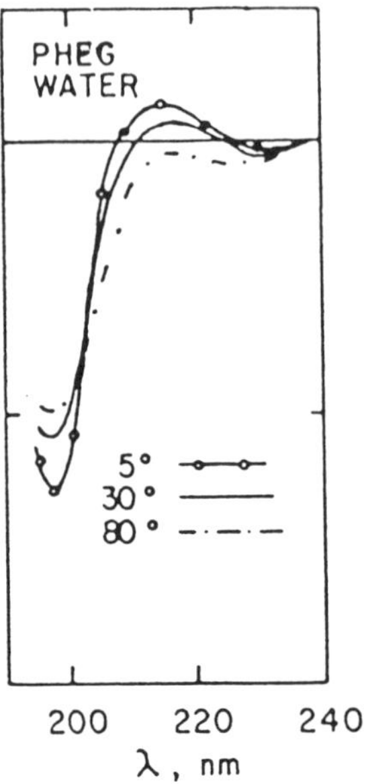

Figure 2. The CD spectrum of poly(N^5-2-hydroxyethyl-Gln) in water as a function of temperature. (Reproduced from Mattice et al., *Macromolecules* **1972**, *5*, 729–734 with permission. Copyright © 1972 American Chemical Society.)

imply an unordered conformation. For example, the CD spectra of bovine pancreatic trypsin inhibitor[16] and of chymotrypsin[17] have strong negative bands just above 200 nm, but only 23[18] and 38[19] %, respectively, of the residues in these proteins are not in α helix, β sheet or β turn regions, according to X-ray diffraction. In the case of bovine pancreatic trypsin inhibitor, recent work[20] has shown that aromatic side chains play a major role in the 205 nm band.

Unordered polypeptides and peptides show more varied behavior in the longer wavelength region, above 200 nm. Ionized poly(Glu)[11] and poly(Lys)[11,21] in aqueous solution have a moderately strong positive band at 218 nm. Some uncharged polypeptides in an unordered conformation also give such a feature (e.g., poly(N^5-2-hydroxyethyl-Gln)[22] (Figure 2). In addition, many oligopeptides in aqueous solution (e.g., oligo (Ala),[23] the P-peptide[24] fragment from bovine pancreatic ribonuclease in water, and the S-[25] and C-[26] peptides in concentrated guanidinium chloride) have a positive band in this region. Other polypeptides, however, have a negative shoulder in the 215–220 nm region. Such systems include poly(Leu-Lys)[27] at low ionic strength (Figure 3), many denatured proteins,[15,28–31] and proteins such as casein[21] and histones[29] at low ionic strength, for which independent evidence indicates chain flexibility or lack of regular structure.

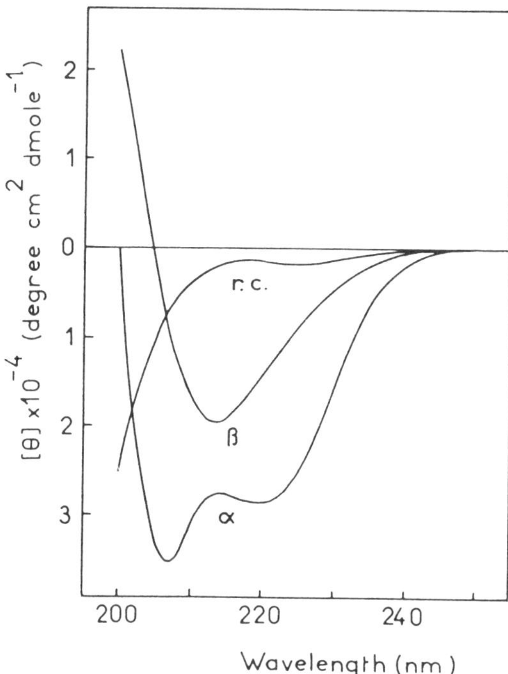

Figure 3. The CD spectrum of poly(Leu-Lys) in water (rc) and in 0.2 M NaCl (β), and of the random copolymer, poly(Leu,Lys) (40% Leu, 60% Lys) in 0.05 M NaClO$_4$ (α). (Reproduced from Brack and Spach, *J. Am. Chem. Soc* **1981**,*103*, 6319–6323 with permission. Copyright © 1981 American Chemical Society.)

A very weak negative CD band in the 235–240 nm region is also observed in the spectra of unordered poly(Glu)[22,32] and poly(Lys)[21,33] under some conditions. There is general agreement that this band is observable at pH values where the side chains begin to be neutralized,[34,35] or if the ionic strength is increased by added salt[36] or the temperature is increased.[37,38] However, there is disagreement on whether the band is present at pH 7 or higher for poly(Glu) and pH 7 or lower for poly(Lys), at low polymer concentrations with no added salt, and at low temperatures (25°C and below). Some groups[6,22,33,34,39,40] have observed such a band under these conditions, while others[36,38,41] have not. The origin of these discrepancies is not clear. Rao and Miller[40] reported that a few percent of residual blocking groups can have a significant effect on the CD spectra of poly(Glu) and poly(Lys). However, the perturbations they observed were in the 220–230 nm region, not the 230–240 nm region. Furthermore, they observed weak negative bands near 235 nm in poly(Lys) and poly(Glu) samples which had no detectable (<0.05%) residual blocking groups.

B. Conflicting Interpretations

The relationship between CD and conformation for unordered polypeptides has been interpreted in two rather different ways. The CD spectrum will depend upon the interactions between neighboring peptide residues, predominantly between nearest neighbors. In the conventional view[23,40,42–44] random coil or unordered peptides and polypeptides exist as an ensemble of an enormous number of different conformers, with the relative geometry of nearest neighbors determined statistically. The nearest neighbor geometries are weighted according to Boltzmann factors obtainable, to a first approximation, from the potential (or better, free) energy as a function of the Ramachandran angles ϕ, ψ. According to this view, there is no such thing as *the* CD curve for the unordered conformation. Changes in temperature, ionic strength, solvent composition, and sequence will alter the relative weights of different regions of Ramachandran space and therefore lead to changes in the CD spectrum.

Tiffany and Krimm[36] have proposed an interpretation which differs in significant respects from the above view, although there are common points. The most significant difference is that Tiffany and Krimm argue that fully ionized poly(Glu) and poly(Lys) at low temperature and low ionic strengths are in an extended helix conformation, similar to the left-handed 3_1 helix of poly(Pro)II (P_{II} helix). Based upon a theoretical treatment[46] of electrostatic interactions, they proposed an interrupted helix model in which left-handed helical segments approximately 4–7 residues in length with 2.5–3 residues per turn, are interspersed with sharp bends. In terms of the conventional view, their model is that one narrow region of the Ramachandran map near that specifying the P_{II} helix becomes overwhelmingly more favorable energetically, and more or less regular helical regions result. They consider the positive band near 215 nm observed in fully ionized poly(Glu) and poly(Lys) as indicative of this extended helix.

In addition to the polyelectrolyte systems, for which an extended helical conformation is plausible, they postulated at least partial extended helical character for poly(N^5-2-hydroxyethyl-Gln),[2,32,38] oligo(Ala)[23] and, at low temperatures, urea- and guanidinium chloride-denatured polypeptides and proteins.[47]

Tiffany and Krimm's[36] proposal was based upon three arguments. (1) Random coil forms of collagen[37] (gelatin) and poly(Pro)[48] (Figure 4) have a single negative CD band with a maximum near 200 nm. (2) The two-band spectra of ionized poly(Glu) and poly(Lys) closely resemble that of poly(Pro)II,[48] allowing for a red shift of ca. 10 nm in the latter, which can be attributed to the red-shifted absorption spectrum[49] of tertiary amides compared with secondary amides. (3) Theoretical calculations[46,50,51] indicated that a polypeptide with charged side chains should adopt an extended helical conformation with about three residues per turn.

Tiffany and Krimm's paradigm for the unordered conformation was poly(Pro) in concentrated $CaCl_2$ solution.[37,48] It has subsequently been demonstrated by both infrared[52] and NMR[53–55] that in concentrated $CaCl_2$ solution, poly(Pro) has a

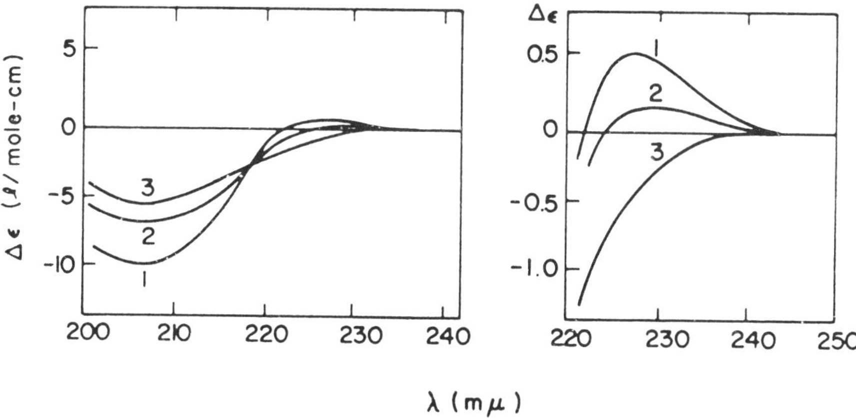

Figure 4. Poly(Pro) in water (curve 1, form II), 3MCaCl₂ (curve 2), and 6MCaCl₂ (curve 3). (Reproduced from Tiffany and Krimm, *Biopolymers* **1968,** 6, 1767–1770 with permission. Copyright © 1968 John Wiley and Sons, Inc.)

substantial fraction (ca. 1/3) of *cis* residues. It is also likely that in denatured collagen a significant fraction of the proline residues are in the *cis* conformation. Therefore, neither of these systems is a good model for typical polypeptide chains with a small fraction of Pro residues at which *cis-trans* isomerization can occur. Furthermore, subsequent CD studies of thermally denatured collagen[56,57] showed that the spectrum is more complex than supposed by Tiffany and Krimm, with a negative tail at long wavelengths in aqueous solution which becomes a distinct shoulder in hexafluoroisopropanol, presumably due to the nπ* transition. Other candidates for unordered polypeptide models proposed by Tiffany and Krimm[37] were various polypeptides and proteins in concentrated salt solutions, in some cases after heating. Although these systems are no doubt unordered, their suitability as models for unordered polypeptides at moderate ionic strengths must be considered questionable.

The electrostatic argument[36] for the P_II-helix-like conformation have also come under attack. The theoretical calculations[46,50,51] which formed the basis for this argument have been criticized on several grounds.[45,58] The calculations started from the assumption of a regular helical conformation and did not take into account the entropic contributions which result from variations in the backbone and side-chain dihedral angles. In addition, a low dielectric constant was assumed and screening by counterions was neglected. Rai and Miller[45] concluded that an extended helix is very unlikely to be more stable than an unordered, statistically determined conformation. However, they acknowledged that locally ordered conformations could be significant at very low ionic strengths.

How did Tiffany and Krimm interpret the CD spectra observed under conditions

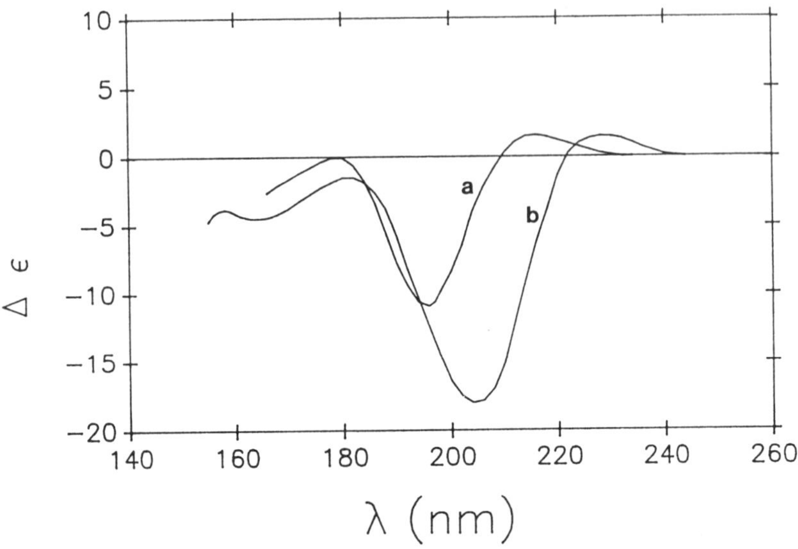

Figure 5. Vacuum uv CD spectra of unordered poly(Glu)[41] and poly(Pro)II[57]. The poly(Glu) spectrum was obtained in water at pH 8, and that of poly(Pro)II in trifluoroethanol. Both spectra were obtained at room temperature.

other than those of high charge density, low ionic strength and low temperature? Their papers are not entirely clear on this point. In their first paper,[36] they attributed the three-band spectrum to an intermediate structural state, without further comments. Their view of the nature of this intermediate state is summarized near the end of a later paper:[37]

> Thus, we believe that while the overall chain structure of PGA at high pH may be considered to be random at the local level the peptide group orientations are correlated over long enough sequences to give a characteristic extended-helix CD spectrum. The term "unordered" is therefore meant to signify an absence of correlation in ϕ, ψ for neighboring peptide groups, thus destroying any specific features of the CD spectrum which arise from any such correlation. It should be noted that intermediate degrees of correlation are possible; this may explain structural features seen in the 210–230 nm region of the CD spectrum in some cases.

Only one of the three arguments presented by Tiffany and Krimm[36] for the extended helix conformation has remained unchallenged—the similarity of the CD spectra of ionized poly(Glu) and poly(Lys) to that of poly(Pro)II. Extension of the CD spectra into the vacuum UV showed that this strong resemblance persists in that region. Figure 5 shows the data of Johnson and coworkers for ionized poly(Glu)[41] and poly(Pro)II.[57] Both polypeptides show a second negative band, about one fourth as intense as the negative band near 200 nm. In poly(Glu), this band is at 175 nm, while in poly(Pro)II it is at 165 nm.

C. Effects of Urea and Guanidinium Chloride

The effects of the classical denaturants urea and guanidinium chloride (GuCl) on the CD of unordered polypeptides and proteins are of interest. Measurement[59,60] of intrinsic viscosity for a series of proteins in concentrated GuCl and urea solutions yielded a dependence on chain length consistent with the proteins being hydrodynamic random coils.[60,61] Tanford concluded from his review of the literature in 1970[62] that proteins denatured by other means generally retain some residual secondary structure.

Tiffany and Krimm[47] studied the effects of GuCl and urea on the CD of poly(Glu) and poly(Lys). In concentrated urea solutions, both of these polypeptides exhibited a more intense 218 nm positive band than in aqueous solution. In GuCl, these polyions showed more complex behavior. At lower concentrations of GuCl, the 218 nm band decreased in amplitude, followed by an increase. In 5.7 M GuCl, the positive ellipticity of poly(Glu) regained only 80% of the intensity observed in water, but poly(Lys) exhibited about 125% of the intensity characteristic of aqueous solution. Both denaturants led to monotonic increases in the long-wavelength positive CD band of poly(Pro). Tiffany and Krimm attributed the increase in the positive CD bands for poly(Lys), poly(Glu) and poly(Pro) at high denaturant concentrations to binding of the denaturants to backbone carbonyl groups, accessible in the relatively open P_{II}-helical conformation, leading to a more rigid conformation. At low concentrations, GuCl causes a decrease in the long-wavelength CD bands of the charged polypeptides in the same way that NaCl does, by shielding the charges on the side chains and decreasing the electrostatic stabilization of the P_{II} helix. At high concentrations, GuCl stabilizes the P_{II} helix by binding to the peptide backbone. In addition to the rigidifying effect invoked by Tiffany and Krimm, it is possible that the guanidinium ion and urea may specifically stabilize the P_{II} helical conformation by participation in hydrogen-bond bridges between peptide groups. Evidence that GuCl and urea bind to peptide groups has been presented by Robinson and Jencks[63] and by Lee and Timasheff.[64]

Privalov et al.[31] compared the CD spectra of several proteins denatured with GuCl, acid and heat. They found that thermally unfolded proteins and those denatured by acid and disulfide reduction had generally similar CD spectra, but that at low temperatures GuCl-denatured proteins had significantly different CD spectra in the long-wavelength region. In all cases, the GuCl-denatured protein had a significantly more positive ellipticity at 222 nm. As the temperature increased, the differences became smaller, with the GuCl-denatured proteins showing the strongest temperature dependence. They suggested that the CD differences and their temperature dependence may be due to GuCl stabilization of the locally extended helical conformation proposed by Tiffany and Krimm.[47] Despite these differences in CD, Privalov et al. neglected the differences in the conformation of the various denatured forms and considered them all to be close to the random coil for the purposes of their thermodynamic analysis. This assumption is supported by the observation[65] that if the thermal contributions of solvation by GuCl are correctly taken into account, the heat capacities of lysozyme in the thermally denatured and GuCl-denatured states are identical.

Privalov et al.[31] also studied a heptapeptide (Glu-Lys-Lys-Leu-Glu-Gln-Ala) and found a significant positive band near 218 nm at 10°C, which increased in intensity upon addition of 6 M GuCl. At 80°C, the long-wavelength CD was negative in aqueous buffer, but was shifted to more positive values upon addition of 6 M GuCl.

D. Low-Temperature CD Studies

Recently, Drake et al.[6] have measured the CD of poly(Lys) in 1,2-ethanediol/water mixtures over the temperature range of ca. −100°C to +80°C (Figure 6). Over this wide temperature range, a tight isodichroic point at 203 nm is observed. The positive band at 218 nm increases by roughly a factor of three upon going from room temperature to −105°C, while the negative band near 195 mn roughly doubles in intensity. On raising the temperature, the 218 nm positive band disappears and at 60°C, the CD is negative over the entire wavelength range of 250–190 nm. Drake et al. also report measurements over a broad temperature range on poly(Lys) in water, 4 M urea, and 5 M NaCl, all at pH 7.6. An isodichroic point is also observed in water over the temperature range of 6°C–87°C. The existence of the isodichroic points implies that the system can be described by two different "states," one of which predominates at low temperatures, and the other of which prevails at high temperatures. The high-temperature form is favored by concentrated NaCl, and in 5 M NaCl the CD spectrum shows only a slight temperature dependence over the range of 7°C–87°C. The CD spectrum in 4 M urea at 23°C is almost identical to that at 6°C in water, implying that urea stabilizes the low-temperature form.

Drake et al.[6] interpret these data, in combination with recent results from VCD to be discussed below, as providing further support for Tiffany and Krimm's mode.[36] They therefore suggest that the low-temperature form is close to that of a P_{II} helix.

E. Vibrational Circular Dichroism

Before discussing the VCD studies of ionized poly(Lys) and poly(Glu), a brief summary of some relevant aspects of the vibrational spectra of the unordered conformation is in order. The ionized form of poly(Glu) and poly(Lys) in D_2O have their amide I' frequencies at 1665 cm^{-1} in Raman and 1646 cm^{-1} in infrared absorption.[66] Chirgadze et al.[67] analyzed the infrared amide I' band of both

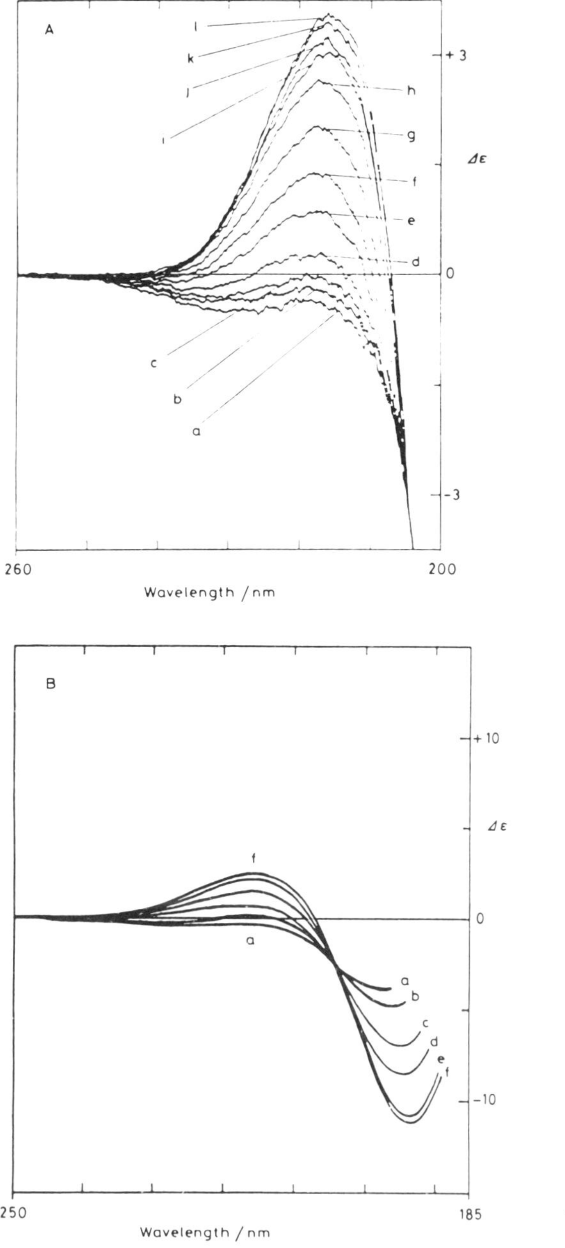

Figure 6. CD spectra of poly(Lys) in 1,2-ethanediol water (2:1), pH 7.6 (A) 260–200 nm region: (a) 82, (b) 69, (c) 60, (d) 45, (e) 21, (f) 2, (g) –26, (h) –55, (i) –75. (j) –85, (k) –95, (l) –105 °C. (B) 250–185 nm region: (a) 84, (b) 60, (c) 22, (d) –28, (e) –8, (f) –102 °C. (Reproduced from Drake et al., *Biophys. Chem.* **1988**, *31*, 143–146 with permission. Copyright © 1988 Elsevier Science Publishers B.V. (Biomedical Division).)

polypeptides and found the major component at 1646 cm^{-1} and a minor component at 1669 cm^{-1}. The large difference between the Raman and infrared amide I' frequencies (ca. 20 cm^{-1}) implies significant coupling between amides, presumably nearest neighbors. This strong coupling is analogous to the difference between absorption and CD maxima in the ultraviolet. Painter and Coleman[68] carried out calculations of the coupling between vibrational transition moments in neighboring peptide groups as a function of ϕ, ψ. Their analysis led them to favor the α helix regions of the Ramachandran map, as these regions give splittings between the infrared- and Raman-allowed vibrational exciton components of 15–20 cm^{-1}, whereas the P$_{II}$ and β sheet regions give smaller splittings, of the order of 5 cm^{-1}. The splitting of the Raman and infrared bands for poly(Pro)II is indeed relatively small, ca. 7 cm^{-1}, based upon the Raman frequency of 1631 cm^{-1} in H$_2$O.[69] and the infrared frequency of 1624 cm^{-1} in D$_2$O.[52] (There should be a negligible difference between the amide I frequency of poly(Pro)II in H$_2$O and D$_2$O because of the absence of exchangeable hydrogens.)

Sengupta and Krimm[70] have reported a Raman spectral study of wet hexagonal crystals of the Ca^{+2} salt of poly(Glu), in combination with a normal mode analysis of helical, charged poly(Glu) conformers. They observed two strong Raman bands in the wet crystals in the C$^{\alpha}$C stretching region (900–1000 cm^{-1}). A band at 928 cm^{-1} persists in the dry state and is assigned to the α-helix conformation. This assignment is based upon the observation of the corresponding band at 924 cm^{-1} in solid, neutral poly(Glu)[71] and X-ray diffraction studies[72] of the Mg^{+2} salt of poly(Glu). The other band, at 943 cm^{-1} is only observed in the wet crystals and has been tentatively assigned to an extended helix on the basis of X-ray diffraction data.[73] In solution, charged poly(Glu) has a Raman band at 948 cm^{-1}.[74,75] The normal mode calculations of Sengupta and Krimm[70] predicted that the C$^{\alpha}$C stretch frequency varies in a nearly linear fashion with the helical rise per residue and with the Ramachandran angle ϕ. Assuming such a linear variation, the observed frequency of 943 cm^{-1} for the putative extended helix implies $\phi \sim -99°$, using the frequencies of α-helical (928 cm^{-1}) and β-sheet[76] (956 cm^{-1}) poly(Glu) as calibration points. This ϕ value is close to that predicted ($-96°$)[51] for an extended helix which minimizes the repulsions between charged glutamate side chains.

These results provide support for the assignment of an extended helix conformation in the wet crystals, although the evidence remains indirect. The similarity of the frequency of charged poly(Glu) in solution adds further support to the Tiffany and Krimm[36] proposal. As Sengupta and Krimm[70] note, the C$^{\alpha}$C stretch band width is about three times larger in solution than in the wet crystal, consistent with a wider distribution of ϕ angles in solution.

Yu et al.[77] have observed the Raman spectrum of poly(Lys) in water at pH 3.7 between 4°C and 60°C. In contrast to the CD spectrum, the Raman spectrum shows no temperature-induced variation over this range. The vibrational frequencies and Raman intensities may simply be much less sensitive to the conformational changes involved than are the CD spectral parameters.

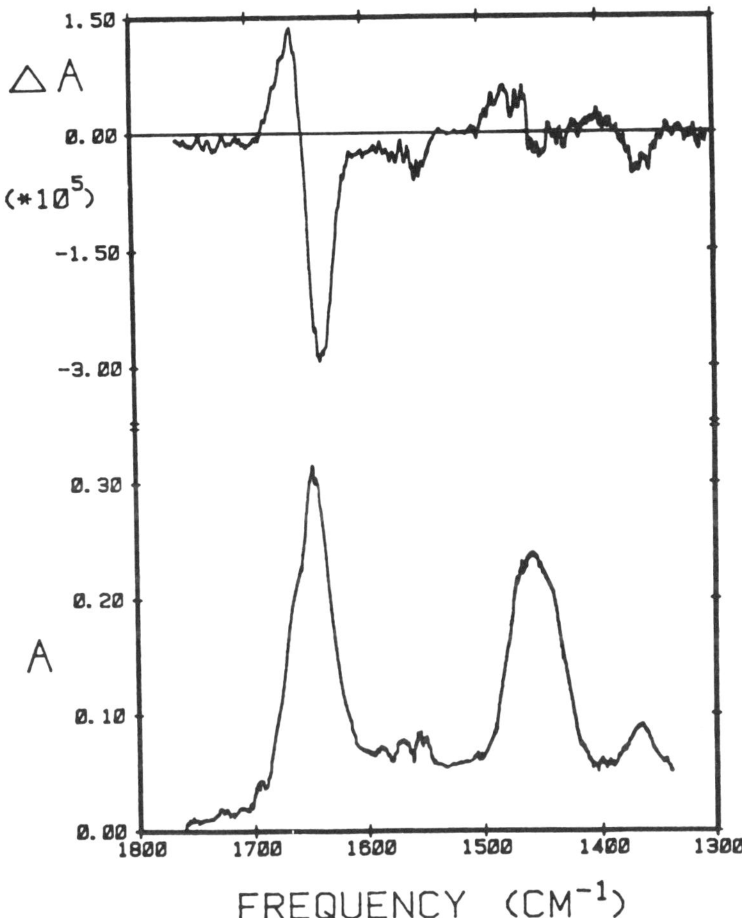

Figure 7. VCD and IR absorption spectra of poly(Lys). (a) Unordered form in D_2O, pH 7.3; (b) α helix in CD_3OD-D_2O (96:4) (Reproduced from Yasui and Keiderling, *J. Am. Chem. Soc.* **1986,** *108*, 5576–5581 with permission. Copyright © 1986 American Chemical Society.)

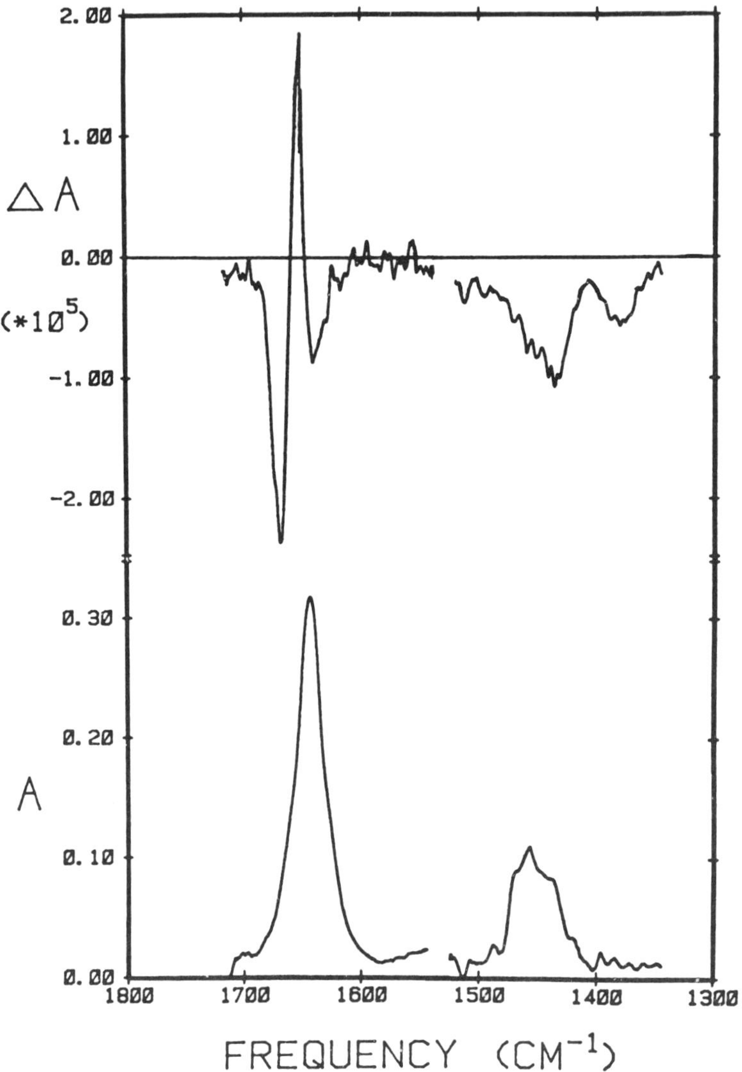

Figure 7. (continued)

VCD studies of poly(Lys) have recently been reported by Yasui and Keiderling[7] and Paterlini et al.[8] For ionized poly(Lys) in D_2O, both groups report that the amide I VCD has a strong negative couplet (Figure 7a), which is opposite in sense to the dominant couplet in the right-handed α-helix (Figure 7b). The value of the peak-to-peak $\Delta A/A$ is comparable to that for the α-helix, reinforcing the inference from electronic CD that the peptide groups in the unordered form are in a distinctly dissymmetric environment. The VCD spectrum is very similar to that observed for poly(Tyr) in dimethylsulfoxide,[78] another unordered polypeptide system.

The strength of the amide I VCD band and the fact that the couplet is opposite in sense to that of the well-characterized right-handed α helix suggested a left-handed helical structure to both groups. In their study of poly(Tyr) in dimethylsulfoxide,[78] Yasui and Keiderling considered the possibility of a left-handed α-helix to account for the VCD. They noted, however, that the amide A VCD band of poly(Tyr) in dimethylsulfoxide is distinctly different from that of poly(AspOBzl) in $CHCl_3$,[79,80] a well-authenticated left-handed α helix. The amide A band cannot be studied in aqueous solution, however. They noted that the amide I band widths in absorption for unordered poly(Lys) were somewhat broader than those for right- or left-handed α helices. Paterlini et al.[8] interpreted the amide I VCD spectrum of ionized poly(Lys) as providing support for the Tiffany and Krimm[36] model, although they noted that until more detailed analysis of the normal modes and transition moment coupling are carried out, an assignment of helix sense cannot be definitive.

Paterlini et al.[8] obtained puzzling VCD results for poly(Lys) in concentrated NaCl solutions. When the poly(Lys) solution was made 5 M in NaCl, the VCD amplitude began to decrease and the amide I absorption band began to broaden and decrease in area. These changes took place over a period of several hours, leading after 4–5 days to a very weak VCD spectrum which, despite a poor signal/noise ratio, qualitatively resembles that of the sample in the absence of salt, but with a $\Delta A/A$ ratio which is less than half as large. When the NaCl was added before a final evaporation to complete H–D exchange, the absorption spectrum showed a small decrease relative to the case where NaCl was added after the D_2O exchange is complete, but there was no discernible VCD signal above the noise level. Dialysis of the sample led to recovery of 70% of the VCD amplitude of the salt-free sample. Paterlini et al.[8] interpreted the apparent disappearance of the VCD signal at high salt concentrations as evidence for a truly unordered conformation of poly(Lys) under these conditions. However, these VCD results differ in two respects from the UV CD observations reported by Tiffany and Krimm[37] for poly(Glu) and poly(Lys) in concentrated salt solutions. First, a notable feature of the observed VCD changes is their slow rate of change. Tiffany and Krimm reported no such slow processes in their UV CD studies. Conformational changes in homopolypeptide chains are generally very fast, except for *cis-trans* isomerization[81] and β-sheet formation.[82] Second, Tiffany and Krimm's[37] high-salt models for completely unordered polypeptide chains still exhibited detectable CD near 200 nm, although the band was generally weaker and broader than that of the salt-free polypeptides. The apparent absence of an amide I VCD band is surprising, even if a truly unordered polypeptide chain has been achieved. It seems likely that high polypeptide concentrations may be playing a role in this observation, but the mechanism is far from clear. Further work is required to provide a full interpretation of this result.

VCD studies of poly(Glu)[83] have also been reported recently. Poly(Glu) at neutral and alkaline pH has a negative couplet in the amide I region, with a shape and amplitude which are similar to those of charged poly(Lys).[7,8]

The initial studies of charged poly(Lys)[7,8] only indirectly implied a similarity to the P_{II} conformation. A left-handed conformation was suggested by the presence of a negative amide I couplet, in contrast to the positive amide I couplet characteristic of the right-handed α helix. The relationship to the poly(Pro)II helix was only inferred because of Tiffany and Krimm's work.[36] Kobrinskaya et al.[84] have reported the amide I VCD of poly(Pro)II, and the shape of the spectrum is very similar to that of unordered poly(Lys)[7,8] and poly(Glu),[83] although the amplitude is about twice as large in the former case.

Surprisingly, the amide I VCD reported[83] for poly(Pro)I in trifluoroethanol is also similar to that of poly(Pro)II in the same solvent, differing only by a factor of ca. 2 in amplitude and a 10 cm^{-1} shift to lower frequencies. Since poly(Pro)I is a right-handed helix, this observation issues a cautionary note concerning too facile an assignment of helix sense from the sign of the amide I VCD couplet. The poly(Pro)I helix contains *cis* amide groups which could have an oppositely signed VCD couplet for a given helix sense. Alternatively, the VCD reported for poly(Pro)I could be affected by *cis-trans* isomerization on the time scale of the experiment. The authors mention that mutarotation occurs during the experiment, but the time course of the VCD spectrum is not discussed in this short report.

Further studies by Dukor et al.[85] have been aimed at characterizing the amide I VCD of oligo(Pro). In agreement with studies of uv CD,[86] only 2–3 residues are required to give the characteristic amide I VCD pattern of poly (Pro)II, although the amplitude does depend on the helix length. This observation is relevant because it assures that even short stretches of P_{II} helix will make contributions to the VCD spectrum of an unordered peptide which qualitatively resemble those of the high-molecular weight polypeptide.

III. P_{II} HELICES IN OTHER SYSTEMS

A. Polypeptides and Glycoproteins

Polypeptides other than poly(Pro) having conformations near the P_{II} helix have CD spectra like that of poly(Pro)II. Poly(Ala-Ala-Gly),[87] for which X-ray diffraction indicates a P_{II} helix in the solid state, has a solution CD spectrum at low temperatures which closely resembles those of poly(Pro)II and ionized poly(Lys).

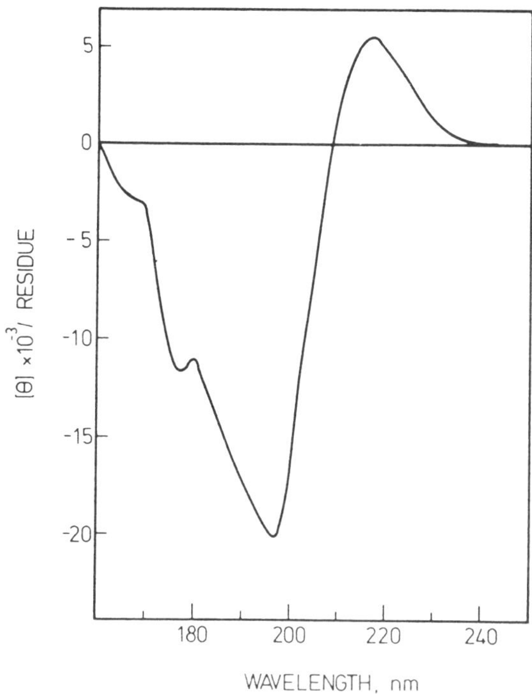

Figure 8. CD spectrum of an antifreeze glycoprotein, corrected for the contribution of the carbohydrate. (Reproduced from Bush et al., *Intl. J. Pept. Protein Res.* **1981**, *17*, 125–129 with permission. Copyright © 1981 Munksgaard.)

Poly(Glu-Ala) at pH 10 and 25°C has a CD spectrum[88] with a positive band at 215 nm, a strong negative band at 196 nm, and a very weak negative band at 238 nm. In both poly(Ala-Ala-Gly) and poly(Glu-Ala), heating leads to a loss of the positive band near 215 nm and a weakening of the negative band near 200 nm. The authors[87,88] interpreted these observations as evidence for a P_{II} helix at low temperatures, melting out with increasing temperature to give an unordered conformation.

Antifreeze glycoproteins from antarctic fish also have a CD spectrum[89,90] like poly(Pro)II, especially after appropriate corrections are made for the contribution of the disaccharides attached to every third residue (Figure 8). The sequence of these proteins is dominated by the repeating tripeptide -Ala-(Gal-β-1→3-GalNAc-α)Thr-Ala. Bush et al.[89] noted that, among all of the known model polypeptides with L-amino acids, only the poly(Pro)II and collagen helices have a positive CD band at long wavelengths. They also noted the similarity to the CD spectrum of poly(Glu) at neutral pH, and Tiffany and Krimm's proposal[36] that this reflects a significant extent of P_{II} helical character. Therefore, Bush et al.[89] proposed that the antifreeze glycoproteins have stretches of 5–10 residues in the P_{II} conformation, interspersed with other conformers so that there is long-range disorder. This latter feature was proposed to retain consistency with quasi-elastic light scattering[91] and ^{13}C NMR[92] evidence for flexible, random-coil-like hydrodynamic behavior.

Subsequent CD, NMR and conformational energy studies[93,94] provide further support for a P_{II} helical conformation for this protein and for a closely related protein from an arctic fish in which half of the Ala residues following the glycosylated Thr are replaced by Pro. The identity of the chemical shifts for a given residue of the repeating tripeptide throughout the chain implies either a three-fold helix or a random coil. The $^{3}J_{\alpha N}$ coupling constants are consistent with ϕ between –80° and –100°, while a strong NOE between N_iH and $C_{i-1}{}^{\alpha}H$ indicates that ψ is positive and therefore argues against conformations in the α_R region.

Synthetic oligopeptides with the sequence [Thr(β-D-Gal)-Ala-Ala]$_n$, with n = 2–7 have recently been studied by CD and NMR.[95] The CD spectra at low temperature (ca. –2°C) in aqueous solution were very similar to those observed by Bush et al.,[89] although the amplitude is larger. There was very little dependence on chain length, with the glycopeptides with n = 2 having nearly the same CD as those with n = 7. At room temperature, the long-wavelength positive band nearly disappeared, while the 200 nm negative band decreased by about 20% in amplitude. At higher temperatures, the 200 nm band continued to decrease in amplitude and the long-wavelength feature became distinctly negative. These changes are all consistent with the behavior of poly(Glu) and poly(Lys) at neutral pH.[6,38,42] Filira et al.[95] interpret their results in terms of a random coil at low temperatures, with small amounts of α helix and/or β sheet at higher temperatures. The authors argue against the P_{II} helical structure proposed by Bush and co-workers,[88,90,93,94] citing the improbability of a P_{II} helix in a hexapeptide, as well as the arguments raised against the Tiffany and Krimm[36] hypothesis by the present author in an earlier review.[2] Filira et al.[95] also point out that the NMR evidence of Bush and coworkers[90,93,94] rules out certain structural types, rather than providing direct evidence for the P_{II} helical structure. This latter comment is valid, but the weight of recent evidence outlined here has led to a re-evaluation of the CD evidence. In addition, the lack of chain-length dependence in the CD of the glycopeptides can be attributed to a combination of the intrinsically low chain-length dependence of CD in oligo(Pro)[86] and the stabilizing influence of the carbohydrate moieties.

A recent review[96] summarizes the evidence supporting the P_{II} helical conformation for antifreeze glycoproteins, including conformational energy calculations of Avanov et. al.[97] These calculations indicate that interactions between the disaccharide groups, which all lie on one side of the helix because of the three-fold symmetry, play a significant role in stabilizing the helical conformation. NOEs between adjacent disaccharide units in the sequence[98] provide evidence for such interactions and thus for the proposed structure.

Plants contain several families of hydroxyproline-rich glycoproteins. The most ubiquitous is extensin, which is an insoluble cell-wall protein. van Holst and Varner[99] have studied a salt-extractable precursor of this cell-wall component by CD. The glycoprotein consists of 35% protein, 62% arabinose and 3% galactose. Of the protein, 45% of the residues are Hyp, to which the arabinose is attached, predominantly as trimers and tetramers. The CD spectrum of this glycoprotein resembles that of poly(Hyp), with the amplitude of the negative band (calculated on the basis of the mean residue weight of the amino acids, and neglecting the sugars) agreeing well with that of the homopolymer. The maximum of this band in the glycoprotein is at 200 nm, as opposed to 206 nm in poly(Hyp), which is attributable to the presence of only 45% tertiary amide bonds in the natural protein, versus 100% in the homopolymer. It was concluded that the protein as isolated is fully in the P_{II} conformation. When the protein was deglycosylated, the spectrum approached that of an unordered conformation, indicating a significant stabilizing role for the oligosaccharide chains in the native protein.

A hydroxproline-rich glycoprotein lectin (agglutinin) has been isolated from potato tubers, and has been the subject of two CD studies.[100,101] Matsumoto et al.[100] concluded that this protein contains 40% β-sheet and 60% α-helix, apparently based upon the molar ellipticity at 195 nm. However, as pointed out by van Holst et al.,[101] Matsumoto et al. calculated the mean residue ellipticity based upon a mean amino acid residue weight which included the carbohydrate, thus leading to $[\theta]$ values which were too small by a factor of ca. 2. Furthermore, the conformational mixture which they proposed cannot account for the positive CD band observed near 225 nm for this protein. The positive band near 225 nm is significantly stronger, relative to the negative band near 200 nm, than in poly(Pro) and poly(Hyp). The sequence of the lectin[102] indicates the presence of two domains, one rich in Hyp and carbohydrate, the other rich in cysteine and heavily cross-linked with disulfides. Cleavage of the disulfides by performic acid or by reduction and alkylation leads to a spectrum which more closely resembles that of poly(Hyp). Quantitative analysis of the CD, using poly(Hyp) as a standard, yielded an estimate of 35% P_{II} conformation, in good agreement with the sequence information, according to which the Hyp-rich domain accounts for 38% of the protein. Deglycosylation leads to a loss of the P_{II} conformation, as with the cell-wall protein,[99] indicating the importance of the oligosaccharide chains in stabilizing this conformation in the native protein.

B. P_{II} Helices in Globular Proteins and Peptide Hormones

Adzhubei and coworkers[103,104] examined the distribution of Ramachandran ϕ, ψ angles[13,105] for the residues in 68 proteins for which structures were available in the Protein Data Bank.[106] They found two maxima in the distribution function

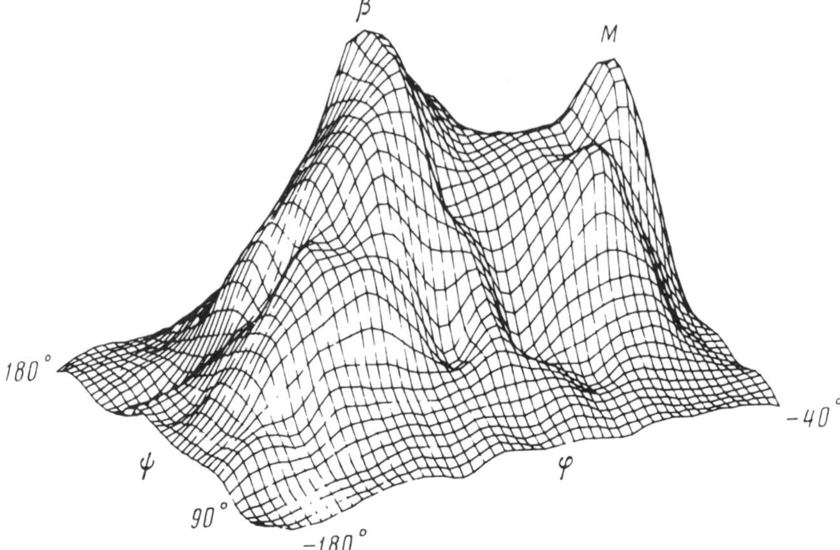

Figure 9. Three-dimensional representation of the distribution of (ϕ,ψ) pairs for amino acid residues in 68 proteins. Only the β/P_{II} region is shown ($-180 < \phi < -40$, $90 < \psi < 180$). (Reproduced from Adzhubei et al., *Biochem. Biophys. Res. Commun.* **1987,** *146,* 934–938 with permission. Copyright © 1987 Academic Press, Inc.)

(Figure 9) in the upper left quadrant of the Ramachandran map, generally referred to as the β region. One of these maxima, at ($-115°$, $+130°$), falls near the classical β-sheet conformations proposed by Pauling and Corey.[107] The other maximum, at ($-65°$, $+140°$), is near the P_{II} conformation. The two overlapping maxima were resolved by eliminating those residues involved in β-sheets, as listed in the Protein Data Bank files. Such residues are involved in interchain hydrogen bonds and are distributed more or less evenly about the classical β-sheet conformation.[107] The remaining residues are less evenly distributed about the second maximum. The residues in this region appear to be hydrogen-bonded to either water or side-chain groups. Adzhubei et al.[103] inferred that the absence of interchain H-bonds will lead to increased chain flexibility in these regions and therefore called this conformation the M (mobile) conformation.

The fraction of residues found in the M conformation was 19.1%, to be compared with 19.6% in β, 52.7% in α_R, 2.5% in the α_L, and 3.0% in the bridging region between the α_R and β conformations. Thus, the frequency of residues in the M conformation is comparable to that in the β conformation. However, the majority (55%) of these residues are isolated residues, while the frequencies of residues which are part of higher order clusters are 22, 12, 8, 2 and 1% for n = 2–6, respectively. Only three examples of six consecutive residues in the M conformation were found among the 68 proteins examined. These were residues 110–115 in an immunoglobulin F_{ab} fragment, 2–7 in avian pancreatic polypeptide, and 59–64 in cytochrome c_{551}. The prevalence of isolated residues is consistent with the absence of inter- or intra-chain hydrogen bonds and the consequent lack of cooperativity.

Two amino acids show a preference for the M region vis-á-vis other regions of the Ramachandran map. These are Pro, for which the frequency of occurrence in the M region is more than twice as large as its average frequency, and Ser, which occurs about 50% more frequently in the M region than expected on average. Cys, Tyr, Thr, Trp, Asn, Arg and Val are slightly enriched in the M region as compared with their average rates of occurrence. The remaining amino acids are less frequently found in the M region than expected on a random basis. The only strikingly low frequency is that for Gly, which is about 2.5 times less frequent in the M region than expected on average.

Makarov et al.[108] have interpreted the CD spectra of several pituitary hormones, including adrenocorticotropic hormone (ACTH), a C-terminal fragment of β-lipotropic hormone (β-LPH), and β-melanocyte-stimulating hormone (β-MSH), as evidence for the presence of the P_{II} conformation in these peptides. They observed a positive CD feature with a maximum at 222–227 nm at temperatures near 0°C, the amplitude of which decreased linearly with temperature. The CD in this region becomes negative above a temperature which depends on the hormone: near 30°C for β-MSH, 15°C for ACTH, and 7°C for the β-LPH fragment. However, the wavelength of this feature, well above 220 nm, is substantially longer than that of the positive features in charged poly(Lys) and poly(Glu), near 217 nm. These peptide hormones have regions which are relatively rich in Pro. For example, β-MSH has 3 Pro out of 18 residues, with two successive Pro residues near the C-terminus. Makarov and Lobachov (personal communication) have pointed out that Pro-rich regions in the P_{II} conformation would be expected to have their positive long-wavelength CD band at longer wavelengths, by analogy with poly(Pro) II, with $\lambda_{max} = 228$ nm. However, this feature might also be due to one or more aromatic amino acids which has a strong positive band in a conformer stable at low temperatures. Makarov et al.[108] argued against this alternative interpretation, citing the pH-independence of the CD. Although the lack of dependence of the peptide hormone CD on pH at alkaline values rules out tyrosine as a candidate, each of the peptides has a single tryptophan for which the CD would be independent of pH at accessible alkalinities. In addition, Makarov et al.[108] pointed to the sensitivity of the observed band to temperature and solvent, contrasting this to the weak dependence of the CD of aromatic amino acids on these variables. Further studies are required to distinguish between these possibilities.

IV. THE P_{II} CONFORMATION AT THE INDIVIDUAL RESIDUE LEVEL

A. Conformational Energy Calculations

Tiffany and Krimm's proposal[36] requires that the P_{II} region of the Ramachandran map is strongly favored in many different systems. With poly(Pro) itself, the constraints imposed by the pyrrolidine ring provide a reasonable explanation for a restricted conformational range. Electrostatic repulsions could be responsible in the case of the polyelectrolytes poly(Glu) and poly(Lys). However, the variety of systems which exhibit a positive 215 nm band makes it necessary to invoke a more general explanation for a strong conformational energy minimum in the poly(Pro)II region.

The question is, what interactions might stabilize the P_{II} conformation in the absence of the constraints of a pyrrolidine ring? The P_{II} helix does not require interactions of longer range than between nearest neighbors for stability, unlike the α helix in which hydrogen bonds between adjacent turns are essential for stability. Therefore, if a P_{II} helix is to be stable, it would be expected to correspond to a significant minimum on the peptide energy surface.

Many theoretical and experimental studies have been directed toward characterizing the simplest models of a polypeptide chain, the N-acetyl, N'-methylamides of the amino acids. These terminally blocked amino acids contain two peptide groups and constitute one amino acid residue. Several conformers are considered to be of significance in these systems corresponding to minima in conformational energy. These conformers are illustrated in Figure 10, and characterized by their ϕ, ψ coordinates in Table 1.

The conformational energy map of Scott and Scheraga[109] for AcAlaNHMe, neglecting solvent, shows a broad low-energy plateau in the region of the β (C_5) and P_{II} conformers. Three minima are predicted: in the C_7^{eq}, α_R, and α_L regions, in order of decreasing depth and breadth. The global minimum is near the C_7^{eq} conformation.

Brant et al.[110] obtained a generally similar map (Figure 11), using a different choice of parameters, but found the global minimum was near the P_{II} conformer. Brant et al.[110] tested their potential map by calculating the characteristic ratio, the ratio of the mean square end-to-end distance to that expected for a freely jointed chain (random walk model). The calculated characteristic ratio for poly(Ala) was 9.3, in good agreement with experimental determinations for several polypeptides in various solvents (not including poly(Ala) for solubility reasons), which gave an average value of 9.5 ± 0.5. When the electrostatic interactions were neglected, equivalent to using an infinite dielectric constant, the α_R region contained the global minimum. The characteristic ratio calculated for this potential map, however, was 2.97, in poor agreement with experiment.

These early calculations utilized only potential energy and neglected solvent.

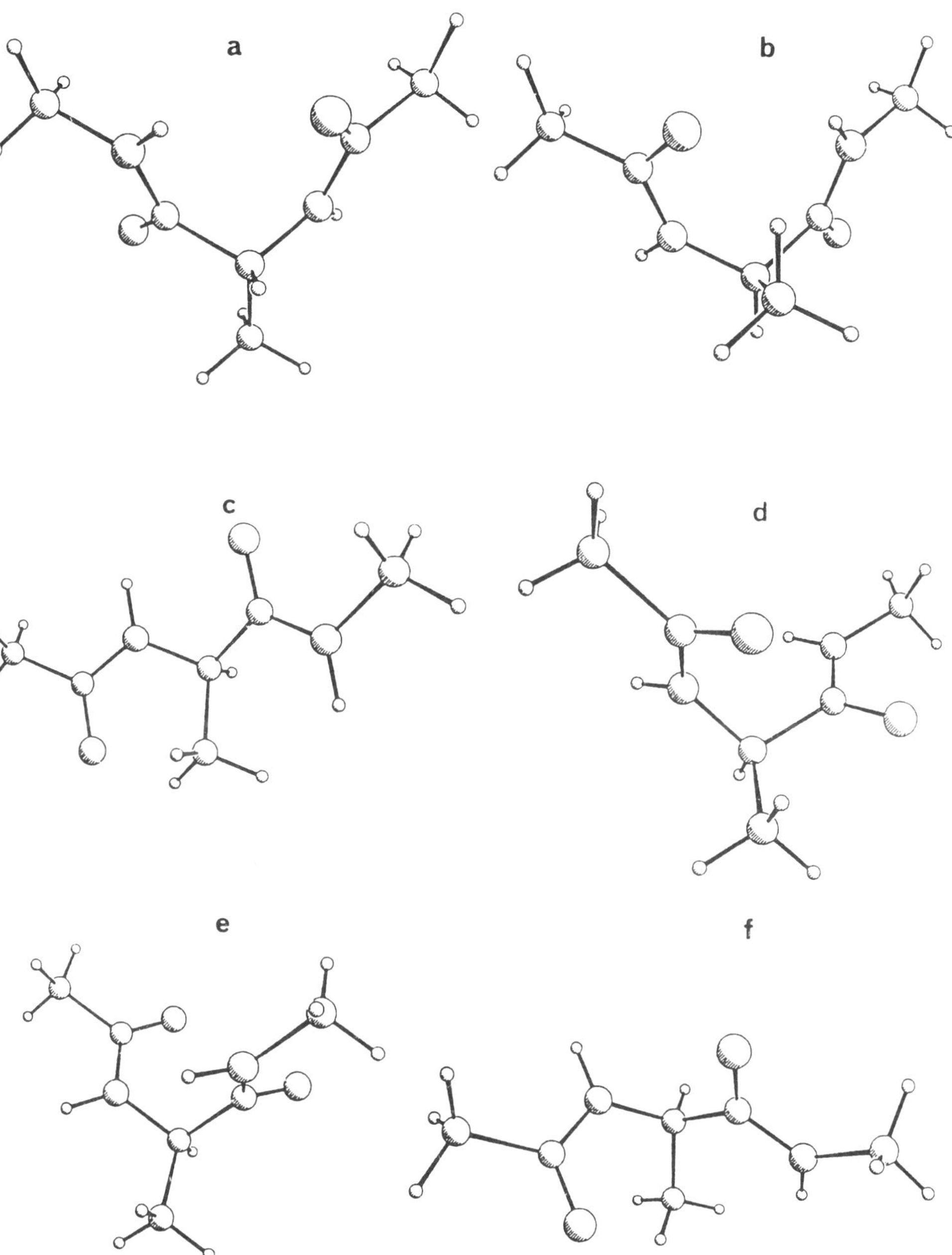

Figure 10. Conformers of AcAlaNHMe. (a) C_7^{eq}, (b) C_7^{ax}, (c) C_5, (d) α_R, (e) α_L, (f) P_{II}. For dihedral angles, see Table 1.

Table 1. Conformers of AcAlaNHMe

Label	ϕ	ψ	Alternative Designations[115]
C_7^{eq}	−80	+80	C
C_7^{ax}	+80	−80	C*
C_5	−150	+150	E
α_R	−60	−60	A
α_L	+60	+60	A*
P_{II}	−80	+150	F

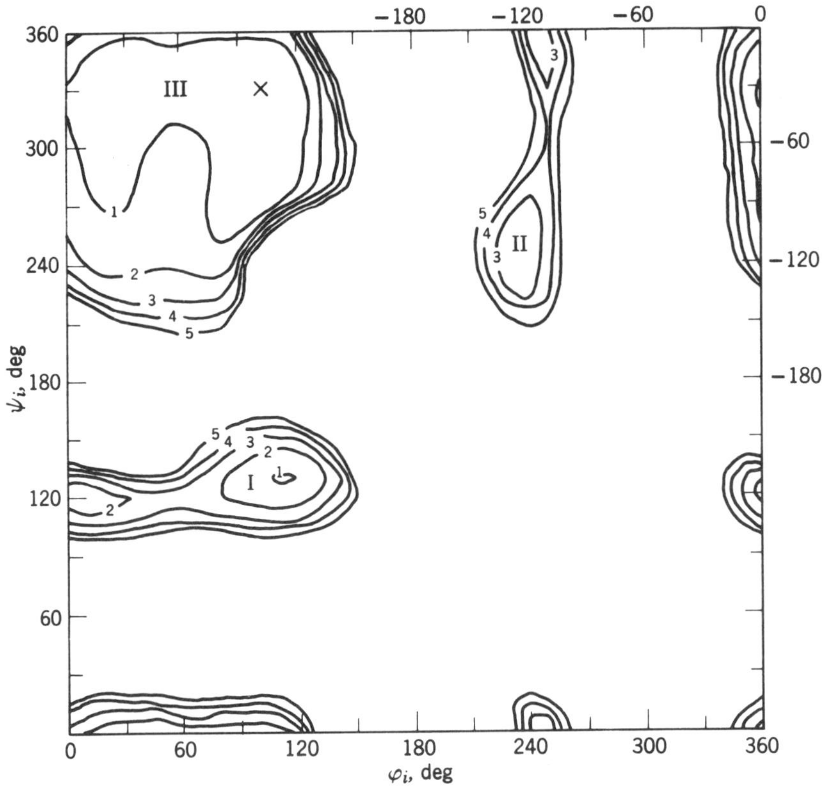

Figure 11. Conformational energy map for AcAlaNHMe. Solvent was not considered, but an effective dielectric constant $\varepsilon = 3.5$ was used. Note that ϕ and ψ as used in this Figure are defined according to an older convention[174] and differ from the current definitions[105] by 180° for each angle. The global minimum is indicated by X, and the contours are at 1 kcal/mol intervals above the global minimum. (Reproduced from Brant et al., *J. Mol. Biol.* **1967**, *23*, 47–65 with permission. Copyright © 1967 Academic Press, Ltd.)

Table 2. Calculated Energies[a] of Unhydrated AcAlaNHMe Conformers[b]

Authors	Methods[c]	C_7^p	C_7^{ax}	C_5	α_R	α_L	P_{II}
Scott and Scheraga[109]	EPF	0	(>4)	(2)	(0.5)	(1)	(1)
Brant et al.[110]	EPF	0	(>5)	(0)	(1)	(2.5)	(0)
Hoffmann and Imamura[111]	EHT	0	(>8)	(−1)	(0)	(2)	(−1)
Maigret et al.[112]	PCILO	0	−0.3	1.4	2.1	(>5)	(0.7)
Momany et al.[113,d]	CNDO/2	0	−0.7	3.1	(>4)	(7)	(4)
Lewis et al.[114]	EPF	0	7.6	0.3	1.9	2.5	
Zimmerman et al.[115]	ECEPP	0	8.8	(>2)	1.1	2.3	(1–2)
Hillier and Robson[116,e]	*ab initio*	0	(9)	(1)	(7)	(8)	(3)
Madison and Kopple[117,f]	EPF	57%		16%	1%	0%	10%
Rossky et al.[118]	EPF	0	2	6	8	(8)	(6)
Gresh et al.[119]	SIBFA						
Roterman et al.[120]	CHARMM[g]	0	2.3	(2)	1.6	(5)	(0.5)
	AMBER[g,h]	0	4.6	1.2	1.0	3.0	(1)
	AMBER[g,i]	0	0.9	1.5	1.4	2.1	1.5
	ECEPP/2	0	7.3	0.7	0.8	2.4	(1.5)
Fermandjian et al.[121]	SIBFA	0	3.9	0.5	7.1	7.9	

[a]Kcal/mol, relative to the C_7^{eq} conformer. Values taken from tabulations when available. Values in parentheses are estimated from potential maps. Blank entries indicate that data for these conformers were unavailable either in tables or figures.

[b]Conformer designations as in Table 1 and Figure 10. However, the exact conformations vary substantially from one calculation to another. In one case, indicated by footnote *d*, a minimum was identified which does not fit any of these categories.

[c]Methods are indicated by the following abbreviations: EPF, empirical force field; EHT, extended Hückel theory[122]; CNDO/2, complete neglect of differential overlap, version 2[123]; PCILO, perturbative configuration interaction with localized orbitals[124]; ECEPP, empirical conformational energy program for peptides[125]; SIBFA, sum of interactions between fragments computed *ab initio*[111]; CHARMM, chemistry at Harvard molecular mechanics[126]; AMBER, assisted model building with energy refinement[127]; ECEPP/2, second version of the empirical conformational energy program for peptides.[128]

[d]The global minimum in this calculation lies 1.0 kcal/mol below the C_7^{eq} conformer and has (ϕ, $\psi = (-10°,+100°)$). This region is not identified with an energy minimum in any other calculations, except that of Pullman et al.,[129] in which it is a local minimum in a region of high energy.

[e]Calculations for *N*-formylalanylamide (i.e., lacking the terminal methyl groups of AcAlaNHMe).

[f]Percent of conformers given instead of energies. Dielectric constant $\varepsilon = 1$.

[g]Using an effective dielectric constant $\varepsilon = 4$. Calculations were also performed with $\varepsilon = 1$ and with $\varepsilon = r$, where *r* is in Å.

[h]Semi-rigid geometry, using harmonic force constants of 400 kcal/mol for ϕ, ψ, bond lengths and bond angles.

[i]Flexible geometry.

In addition, they used fixed bond lengths and bond angles, permitting torsion about single bonds in the backbone and side chains. More recent studies have addressed each of these limitations. In the remainder of this section, this work will be surveyed, with emphasis on the question of the significance and stabilizing influences of the P_{II} conformer.

Table 2 summarizes the results of conformational energy calculations on AcAlaNHMe in the absence of a solvent. While differing considerably in detail, there is general (though not universal) agreement on several points. (1) The C_7^{eq}

conformer is the lowest energy conformer, except in two cases in which it lies at slightly higher energy than the C_7^{ax} conformer. (2) Most calculations place the α_R, P_{II}, and C_5 conformers 1–2 kcal/mol above the C_7^{eq} conformer. (3) The α_L conformer generally lies at least 1 kcal/mol higher in energy than the α_R. (4) Most calculations place the C_7^{ax} conformer ca. 5 or more kcal/mol higher than the C_7^{eq}. However, empirical force field calculations with flexible geometry (CHARMM,[126] AMBER[127]) indicate a smaller energy difference, and semi-empirical MO calculations actually predict that C_7^{ax} is slightly more stable than C_7^{eq}. Overall, these calculations indicate that C_7^{eq} should be the dominant conformer in nonpolar solvents, with smaller contributions from α_R, P_{II}, and C_5 conformers.

Hodes et al.[130] introduced modifications in the ECEPP program[125] to account for free-energy contributions of both specific hydration, corresponding to hydrogen bonding of water molecules with the peptide groups, and nonspecific hydration, with the latter treated as a continuum hydration shell. For AcAlaNHMe, they found that hydration led to an energy minimum in the P_{II} region (F region in their nomenclature), whereas the conformational energy map for the unhydrated molecule[115] showed no minimum in this region. The global minimum was predicted to be in the classical β (their E) region. Using Boltzmann factors calculated from their conformational energy map, the fraction of AcAlaNHMe in the P_{II} conformer was predicted to be 0.16 in water, as opposed to 0.0 in vacuo. Hodes et al. carried out such calculations for all 20 protein amino acids. Their results show that, not surprisingly, AcProNHMe has the highest fraction of P_{II} conformer in both hydrated and unhydrated conditions, increasing from 0.19 to 0.54 upon hydration. Among the other blocked amino acids, AcTrpNHMe and AcArgNHMe were predicted to have P_{II} contents intermediate between AcProNHMe and AcAlaNHMe. Overall, the work of Hodes et al. led to the conclusions that hydration favors the α_R and P_{II} conformers at the expense of the C_7^{eq} conformer, which is the major conformer in vacuo, but which is destabilized by hydration.

The effects of hydration on the conformational preferences of a number of terminally blocked dipeptides were studied by Hodes et al.[131] In most cases, the conformational distribution of each residue in a dipeptide resembles that of the corresponding blocked amino acid. Thus, hydration also favors the α_R and P_{II} conformers in these dipeptides. Of the 23 dipeptides studied, the lowest energy conformer in aqueous solution was predicted to have one residue in the P_{II} region in seven cases and, in two cases (AcProProNHMe and AcAspAlaNHMe), both residues were in this region. Only four unhydrated peptides had a single residue in the P_{II} region, and none had both residues.

The free energies of aqueous AcAlaNHMe in various conformations were compared by Mezei et al.,[132] using a Monte Carlo method. Four conformations were considered: C_7^{eq}, C_5, α_R, and P_{II}. They found that in aqueous solution the α_R and P_{II} conformers are stabilized relative to C_7^{eq}, while C_5 is destabilized substantially. The α_R conformation is predicted to be 0.4 kcal/mol more stable than P_{II}, although the estimated statistical errors are also of this magnitude. Mezei et al.[132] investigated the sources of these differential stabilizations by partitioning the internal energy difference into contributions from solute-water and solute-solute interactions. Solute-water interactions actually favor P_{II} over α_R, but reorganization of the water is energetically unfavorable for PII relative to C_7^{eq}, but enhances the relative stability of the α_R conformer. The water-water reorganization energy is even more unfavorable for the C_5 conformer relative to C_7^{eq}, and it is this factor which is predicted to make the C_5 conformer negligible in aqueous solution. About one more water coordinates the carbonyl oxygen in the P_{II} conformer than in the C_7^{eq} conformer, while the carbonyl of the α_R conformer has about half an additional water of coordination relative to C_7^{eq}. This extra degree of coordination is clearly important, but its energy contribution is impossible to quantitate.

Pettitt and Karplus[133] have used an integral equation method to calculate the potential of mean force for various points on the Ramachandran map of AcAlaNHMe in aqueous solution. (Some aspects of this calculation were corrected in a subsequent paper.[134] This corrected version will be quoted here where the two papers differ.) Comparison of this map with the in vacuo map gives the free energy of solvation for each conformer. The in vacuo potential map shows only three minima: at C_7^{eq}, C_7^{ax}, and C_5. The two C_7 conformers are within 0.1 kcal/mol, with C_7^{eq} lowest, whereas the C_5 conformer lies 3.6 kcal/mol higher. In aqueous solution, C_7^{ax} and P_{II} become the lowest in energy, 0.5 kcal/mol below C_7^{eq}. Surprisingly, this calculation places the α_R conformer at the highest energy, 1.3 kcal/mol above the C_7^{eq} conformation, with the C_5 and α_L conformations at –0.4 and 0.5 kcal/mol, respectively. The solvation energies are predicted to be largest for the P_{II} and α_L conformers, –11.2 and –10.6 kcal/mol, respectively, and only –9.9 kcal/mol for the α_R. Fractional populations predicted by integration over a square area $\pm 20°$ from the energy minima are 12.0, 7.9, 5.7, 2.9, 2.3, and 0.8% for the C_5, P_{II}, C_7^{ax}, C_7^{eq}, α_L, and α_R, respectively.

Anderson and Hermans[135] have used molecular dynamics simulations to calculate the free energy map for AcAlaNHMe in water, and from this the probability distribution map shown in Figure 12. This latter map shows a strong maximum near the classical β conformation[107] ($\phi = -110°$, $\psi = 120°$), with a considerably smaller maximum near the α_R conformation, and still weaker peaks in the C_7^{ax} and α_L regions. Because of the breadth of the maximum in the β region, P_{II} conformations ($-80°$, $+150°$) have a significant probability, but there is no suggestion that they should be dominant.

In Table 3, the results from the recent theoretical studies of AcAlaNHMe aqueous solution are compiled. As in the case of the calculations on the unhydrated molecule, these results show considerable variability. However, most calculations indicate that the C_7^{eq} conformer is not the most stable conformer in aqueous solution.

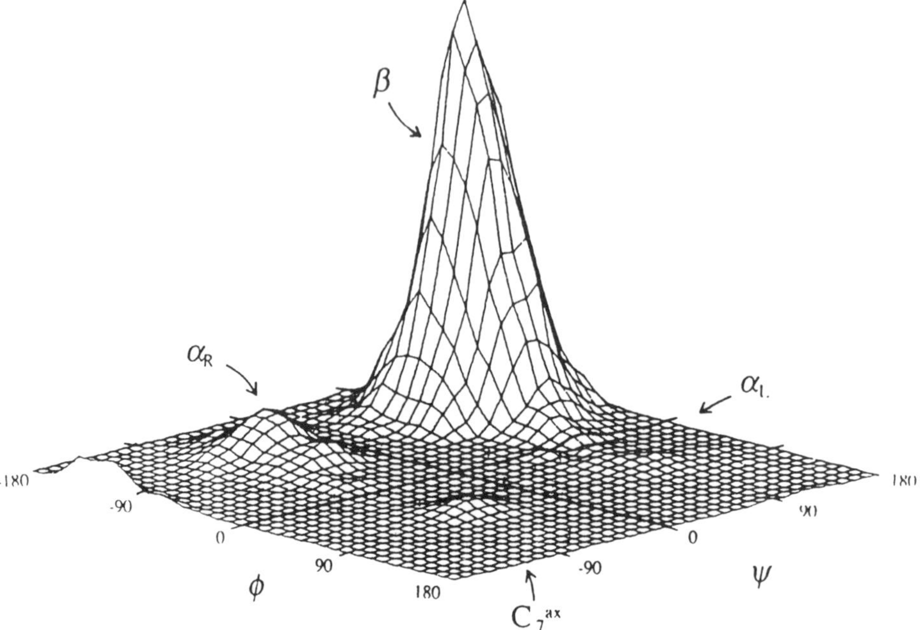

Figure 12. Conformational probability map for AcAlaNHMe in water. The distance from the Ramachandran plane represents the probability for each ϕ, ψ point on the grid. The probabilities were calculated from free energies generated by molecular dynamics simulation of the dipeptide plus 76 water molecules (two molecular layers around the peptide) in a periodic box. (Reproduced from Anderson and Hermans, *Proteins* **1988**, *3*, 262–265 with permission. Copyright ©1988 Alan R. Liss, Inc.)

Instead, the α_R, P_{II}, and C_5 are generally found to be more stable, because hydration stabilizes these conformers more strongly than the C_7^{eq}. The P_{II} conformer is not predicted to be the most stable by any of the calculations, but it is generally only slightly less favorable than either the α_R or C_5 conformers.

B. Rotational Strength Calculations

Bayley et al.[143] carried out a systematic survey of the Ramachandran plane, calculating the $n\pi^*$ and $\pi\pi^*$ rotational strengths for ϕ, ψ pairs on a 20° grid (Figure 13). Their work provides a useful guide to interpreting the CD of peptides. They found that the $n\pi^*$ transition in blocked amino acids will have negative rotational strengths over most of the fully allowed regions, except for a triangle in the upper right of the β/P_{II} region. Their calculations also predicted a positive couplet for the $\pi\pi^*$ transition in the α_R region and the upper right diagonal of the β/P_{II} region, with

Table 3. Calculated Energies[a] of Hydrated AcAlaNHMe Conformers[b]

Authors	Methods[c]	C_7^{eq}	C_7^{ax}	C_5	α_R	α_L	P_{II}
Renugopalakrishnan et al.[136,d]	RF	0	−7.3	(−3)	(−7)	(−7)	(>−2)
Hodes et al.[130]	ECEPP+H	0		−0.2	0.2	1.7	0
Madison and Kopple[117,e]	$\varepsilon = \infty$	15%		21%	12%	1%	24%
	SD	4%		11%	50%	2%	14%
	BO	1%		0%	88%	2%	10%
	N-H[f]	9%		16%	2%	0%	27%
	C=O	5%		15%	10%	0%	54%
	SS	11%		40%	3%	0%	15%
Mezei et al.[132]	MC	0			−3.6		−3.2
Pettitt et al.[137]	IE	0			−5.2		−4.7
Lau and Pettitt[134]	IE	0	−0.5	−0.4	1.3	0.5	−0.5
Anderson and Hermans[135,g]	MD		2.7	0	1.5	2.5	
Grant et al.[138]	AIRF[h]	0	10.3	−3.1	−1.9	0.6	−2.6
	AIRF[i]	0	12.4	−2.4	−4.1	−0.03	−2.1

[a]Kcal/mol, relative to the C_7^{eq} conformer unless otherwise noted. Values taken from tabulations, where available. Other values, estimated from potential maps, are in parentheses. Blank entries indicate that data for these conformers were unavailable either in tables or figures.

[b]Conformer designations as in Table 1 and Figure 10. However, the exact conformations vary substantially from one calculation to another.

[c]Methods are indicated by the following abbreviations: RF, reaction field; ECEPP+H, empirical conformational energy program for peptides[125] plus hydration; MC, Monte Carlo; IE, integral equation method; MD, molecular dynamics; AIRF, ab initio + reaction field; $\varepsilon = \infty$ uses a standard empirical potential field (EPF) with an infinite dielectric constant (i.e., no electrostatic contributions); SD, solvated dipole, includes interaction of peptide dipole located at the center of a spherical cavity with the solvent modeled as a continuum; BO, big oxygen, uses EPF with $\varepsilon = \infty$ but increases the value of the van der Waals radius of O by 1 Å to emphasize conformers in which water can approach carbonyl oxygen; N–H, EPF with oxygen atom 2.8 Å from N of each N–H, collinear with NH; C=O, EPF with a water molecule H-bonded to each carbonyl oxygen; SS, solvent shell model of Forsythe and Hopfinger.[139]

[d]The global minimum was located at ca. (−120°, −60°) with an energy of −9.3 relative to the C_7^{eq}.

[e]Percent of each conformer was given, instead of energies.

[f]In this model, the dominant conformer was a two-fold helical conformer at (−150°, 80°), accounting for 44%.

[g]The global energy minimum was located in the β region (−110°, +120°), which we have equated here to the C_5 conformation. The usual α_R conformation corresponds to a weaker minimum than one at (−120°, −40°). In this case, there is not enough information to relate the conformer energies to that of the C_7^{eq} conformer, which does not appear as a discrete minimum (or at least not as a discrete probability maximum).

[h]Using cavity radii of Rashin and Namboodiri,[140] and a 6–31 G* basis set.

[i]Using the cavity radii based upon Aguilar et al.[141,142] and a 6–31 G* basis set.

negative couplets in the lower left β/P_{II} regions. Bayley et al.[143] pointed out that, assuming the positive band at 218 nm band of poly(Glu) and poly(Lys) is due to the $n\pi^*$ transition, their results would suggest significant contributions from conformations in the P_{II} region of the Ramachandran map.

Nielsen and Schellman[144] reported ORD spectra for several diamides, including AcAlaNHMe, and calculated CD spectra from the ORD by the Kronig-Kramers transformation.[145] By curve resolution, they found a weak positive (−0.017 Deby-Bohr magneton, DBM = 0.9273 × 10^{-38} cgs unit) at 210 nm in water, and a strong negative $n\pi^*$ rotational strength (−0.102 DBM) in dioxane at 227 nm. A couplet was observed for the $\pi\pi^*$ transition in both solvents, but the sign was opposite,

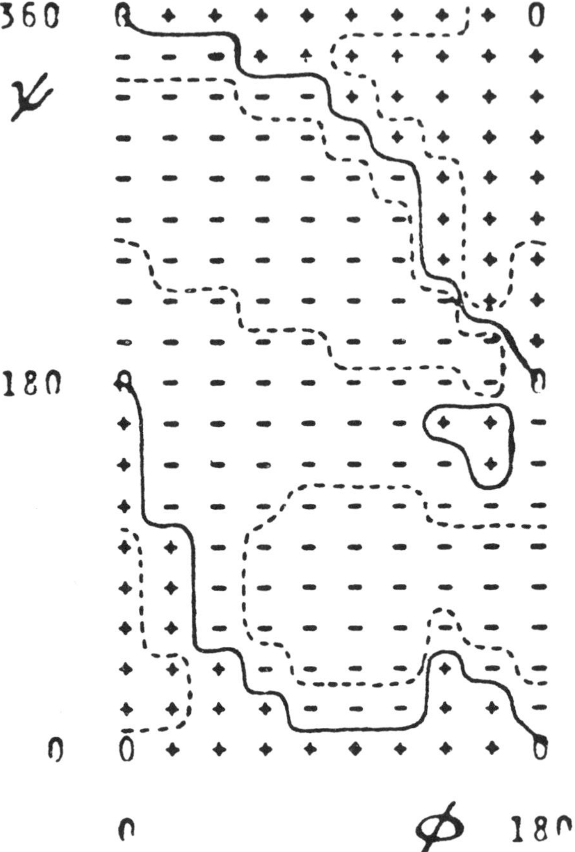

Figure 13.   Map of nπ* rotational strengths for AcAlaNHMe as a function of φ and ψ. The points for which calculations were performed are at 20° intervals. Only the left half of the map is shown. The plus and minus signs indicate the sign of the nπ* rotational strength. The solid lines represent nodes at which the rotational strength vanishes. The dotted lines indicate the boundaries of regions for which the rotational strength has a magnitude greater than 0.1 DBM. Note that an older convention[174] for φ and ψ was used in this figure. The values shown here can be converted to the present convention[105] by subtracting 180° from each angle. (Reproduced from Bayley et al., *J. Phys. Chem.* **1969**, *73*, 228–243 with permission. Copyright © 1969 American Chemical Society.)

being negative in water and positive in dioxane. Nielsen and Schellman carried out rotational strength calculations for some of the other diamides, all of which involved cyclic structures in which either φ or ψ were fixed. They did not report calculations for AcAlaNHMe, but discussed the observed nπ* rotational strengths in terms of the calculations of Bayley et al.[143] They noted that the positive nπ* rotational strength in water suggested a high population in the P_{II} region of conformational space. The negative $n\pi^*$ rotational strength observed in dioxane, as they noted, is consistent with a large fraction of conformational space, including the α_R, $\beta(C_5)$, and C_7^{eq} regions. Of these, they suggested that the α_R conformation is unlikely in nonpolar solvents because of the destabilizing effects of coulombic interactions at the dimer level.

The results of Bayley et al.[143] for the $\pi\pi^*$ transition do not lead to such clear predictions, however. For both the P_{II} and the α_R conformations, exciton coupling in the dimer is predicted to lead to a positive couplet for the $\pi\pi^*$ transition, while in the β conformation, the $\pi\pi^*$ couplet is predicted to vanish. Another conclusion, implicit in the results of Bayley et al.,[143] is that calculations on the $\pi\pi^*$ CD in amide dimers are seriously complicated by the presence of nodal lines running through all three populated regions of the Ramachandran map: α_R, β, and P_{II}. At these nodal lines, the $\pi\pi^*$ exciton contribution changes sign. Thus, the weighting of different parts of the Ramachandran map is crucial.

Tonelli[146] calculated the $n\pi^*$ and $\pi\pi^*$ rotational strengths of amide dimers as a model for the unordered conformation. He carried out calculations for ϕ, ψ pairs on a 10° grid, weighting the calculated rotational strengths by Boltzmann factors calculated from the potential energy map of Brant et al.[110] The results were sensitive to the choice of electronic parameters, but both sets used gave qualitatively satisfactory results, predicting a positive $n\pi^*$ and a negative $\pi\pi^*$ rotational strength.

Ronish and Krimm[147] also used a dimer model for the unordered form, obtaining quite different results from those of Tonelli.[146] They did not give any details of the conformational energy map which they used to obtain their Boltzmann factors, quoting only a private communication. However, they stated that similar results were obtained with the map of Brant et al.[110] used by Tonelli. In addition, their electronic parameters were similar, but not identical to one of the sets used by Tonelli. Ronish and Krimm predicted a moderately strong negative band at 213 nm and a somewhat stronger positive band near 190. They associated the 213 nm negative band with the negative band observed by Tiffany and Krimm[37,48] near 200 nm for various polypeptides in concentrated salt solutions, and considered the absence of a long-wavelength positive feature in their calculated spectrum as evidence for the Tiffany and Krimm[26] proposal that this feature is diagnostic of the P_{II} conformation. However, the large wavelength difference between their calculated negative band and that observed for the putative unordered conformation is disturbing. Although they did not characterize the predicted 213 nm band, it seems likely that it was the $n\pi^*$ band, since they chose 212 nm as the unperturbed wavelength for the $n\pi^*$ transition and 190 nm for the $\pi\pi^*$ transition. The experimentally observed band near 200 nm must be the $\pi\pi^*$ transition, both because of its short wavelength and because of the observation[54,55] of a long tail which is due to the $n\pi^*$ transition.

Two calculations[117,148] have more recently been reported in which the objective was to calculate the CD of several blocked amino acids, including AcAlaNHMe.

Madison and Kopple[117] combined conformational energy calculations with NMR and CD data to explore the conformational equilibria of AcAlaNHMe and AcProNHMe in water, and both aprotic polar and nonpolar solvents. The conformational energies were evaluated by several methods which model the effect of solvation. The CD of AcProNHMe in $CHCl_3$ and cyclohexane shows a strong negative $n\pi^*$ band near 225 nm, which is diminished as one goes to dioxane and acetonitrile, and is not observed as a discrete band in water and alcohols. Calculation of the $n\pi^*$ rotational strength of AcProNHMe as a function of ψ at $\phi = -53°$ yields a large negative rotational strength only for ψ near $+50°$, near the C_7^{eq} conformation. Thus, the CD spectra suggest that the C_7^{eq} conformation is important in nonpolar solvents, but becomes decreasingly so with increasing solvent polarity. This interpretation is strongly supported by NMR measurements. Siemion et al.[149] have demonstrated a linear correlation between ψ for a Pro residue and the difference in the ^{13}C chemical shifts of the β and γ carbons, $\Delta\delta_{\beta\gamma}$. A plot of $\Delta\delta_{\beta\gamma}$ against the CD at 225 nm gives a straight line for the five solvents, ranging from $CHCl_3$ to water, in which both CD and NMR measurements were made. These results, together with NOE measurements, support the predictions of conformational energy calculations omitting the solvent which predict that 88% of the peptide is in the C_7^{eq} conformation.

For the more polar solvents, the CD and NMR data indicate decreased amounts of the C_7^{eq} conformer, and increasing amounts of the α_R and P_{II} conformers. However, these latter two conformers are difficult to distinguish by NMR criteria and, although they differ in the predicted sign of the $n\pi^*$ rotational strength (negative for the α_R and positive for the P_{II}), without a knowledge of the magnitudes for the two forms one cannot determine them quantitatively. However, the near-zero value for the CD at 225 nm in water suggests a substantial but not dominant amount of the P_{II} conformer. The conformational energy calculations which attempt to model the effects of polar solvents predict that solvation sharply decreases the C_7^{eq} conformer population and increases the fraction of molecules in the α_R and P_{II} conformation, with the relative amounts of these two conformers depending on the solvent model. Two of the models predict a 2:1 or a 10:1 preponderance of α_R over P_{II}, but a third model, which only considers solvation of the peptide NH, gives 100% P_{II}. This model should be especially suitable for dimethylsulfoxide, and evidence from NOEs does suggest a low α_R content in this solvent. Unfortunately, CD measurements could not be made in dimethylsulfoxide because of strong solvent absorption below 250 nm.

CD measurements on AcAlaNHMe in various solvents gave results similar to those for AcProNHMe, except that the $n\pi^*$ rotational strength is small and positive for the blocked Ala peptide. The combined CD, NMR, and conformational energy calculations present a picture which closely resembles that for AcProNHMe and gives satisfactory agreement with experiment. In both blocked peptides, the C_7^{eq} conformer is dominant in non-polar solvents, whereas the α_R and P_{II} conformers are prevalent in water, although the relative weights for these two conformers remain uncertain.

The blocked peptides of Ala and Ser, AcAlaNHMe and AcSerNHMe, have been investigated by Dungan and Hooker.[148] We shall consider only the results for AcAlaNHMe. The experimental CD spectra determined by Dungan and Hooker in various solvents agree well with those reported by Madison and Kopple,[117] except that the long-wavelength region did not show a weak positive feature in aqueous solution, but remained negative. On the other hand, measurements on an ethanol solution at $-48°C$ showed a distinct positive feature, with a maximum at 220 nm. Dungan and Hooker carried out theoretical calculations of the CD of AcAlaNHMe for ϕ, ψ pairs covering the Ramachandran map, using a modification of the method of Bayley et al.[143] They displayed their results so that one can readily ascertain the sign of both the $n\pi^*$ rotational strength and the $\pi\pi^*$ couplet (Figure 14). Dungan and Hooker also calculated CD spectra by weighting the results of their rotational strength calculations by Boltzmann factors calculated from the semi-empirical potential function which they used, as well as the results of Lewis et al.[114] and of Zimmerman.[115] They used the same Boltzmann factors for calculations on both polar and nonpolar solutions, but used different wavelengths for the $n\pi^*$ and $\pi\pi^*$ transitions in nonpolar (220 and 186 nm, respectively) and polar (212 and 188 mn). The spectra which they predicted for both solvent types qualitatively resembled those observed in nonpolar solution, with a relatively strong, negative $n\pi^*$ band and a positive $\pi\pi^*$ couplet. Since the potential energy surfaces which they used did not take solvent into account, their inability to reproduce the CD in a polar solvent is not surprising. The dominant conformation in their calculations would be the C_7^{eq} conformation, which has exactly the observed CD characteristics. It would be of interest to repeat their calculations using one of the more recent energy surfaces which take into account the effect of hydration (Table 3), which would decrease the weight of the C_7^{eq} conformer and increase that of the P_{II}, α_R and C_5 conformers.

C. Other Experimental Results

The principal experimental methods, in addition to CD, which have been applied to the elucidation of the conformation of blocked amino acids in solution are IR and NMR. IR studies have focussed on the NH stretching band in nonpolar solvents such as CCl_4 and $CHCl_3$. Such studies[150–159] have indicated the presence of both C_7 and C_5 conformers in these nonpolar solvents. Mizushima and coworkers[150,151] first proposed the C_7 conformer to explain the multiple NH stretch frequencies they observed for AcAlaNHMe in CCl_4. Portnova et al.[152] have estimated that 70% of the NH groups in the blocked dipeptide Ac-D-Ala-D-AlaOMe are involved in C_7 conformers in CCl_4. Bystrov et al.[153] have investigated this dipeptide in 9:1 CCl_4-$CHCl_3$ mixtures and observed 50% of the folded conformation. Avignon and co-workers[151–156] assigned two NH stretching bands in blocked amino acids to intramolecular H-bonded conformers, and thus inferred the coexistence of C_7 and

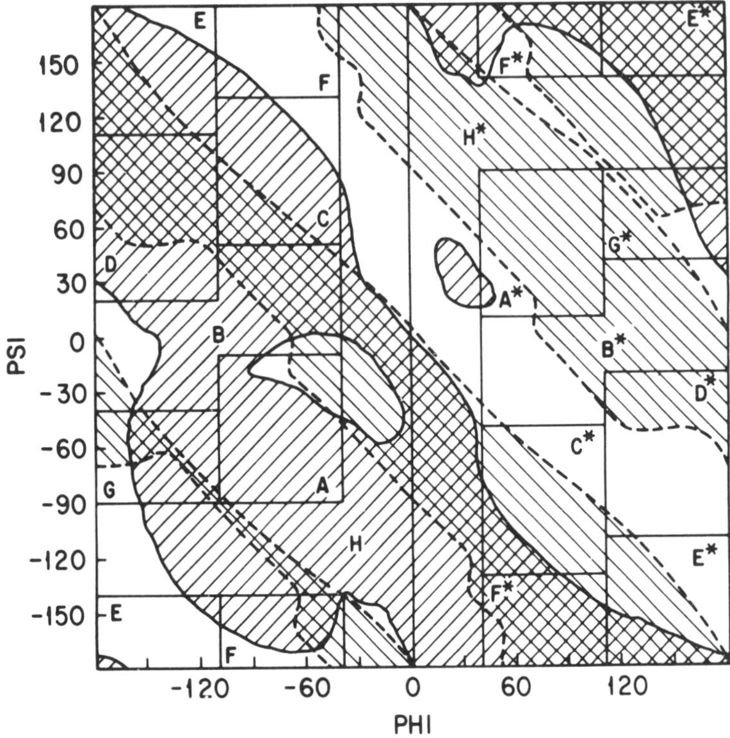

Figure 14. Map of $n\pi^*$ and $\pi\pi^*$ rotational strengths for AcAlaNHMe as a function of ϕ and ψ. The wavelengths assumed for the unperturbed $n\pi^*$ and $\pi\pi^*$ transitions in the monomer were 212 and 188 nm, respectively, values appropriate for aqueous solution. The solid lines represent nodal lines for the $n\pi^*$ transition, and the dashed lines for the $\pi\pi^*$ transition. Shading with positive slope indicates negative $n\pi^*$ rotational strengths, whereas lines of negative slope indicate negative $\pi\pi^*$ couplets. The rectangular regions represent the conformational regions defined by Zimmerman et al.[115] (Reproduced from Dungan and Hooker, *Macromolecules* **1981**, *14*, 1812–1822 with permission. Copyright © 1981 American Chemical Society).

C_5 conformers in CCl_4 and $CHCl_3$. Avignon and Huong[156] estimated that only the C_5 and C_7 conformers of AcAlaNHEt are present in CCl_4 and they are in a 1:3 ratio. Avignon et al.[154] also argued that the NH frequencies for the C_7 conformer are consistent with a C_7^{eq} conformer and not with a C_7^{ax}.

Burgess and Scheraga[157] suggested a reinterpretation of the NH stretching frequencies, which assigned one band to conformers having $\phi = -60°$, which includes the C_7^{eq} and P_{II} conformers, the latter of which they referred to as the γ conformer. They proposed that a band at 3440 cm^{-1}, previously assigned[154–156] to the free NH of the blocked amino acid in the C_7 conformation, or to the weakly

H-bonded NH proton of the C_5 conformer,[151] could also be characteristic of the non-hydrogen-bonded NH protons of the P_{II} conformer. This suggestion was based upon several arguments, one of which was the observation[158] of a single NH stretching band at 3440 cm^{-1} in the spectrum of AcIleNHMe in CCl_4, for which the empirical energy calculations of Lewis et al.[114] predicted the P_{II} conformer to be significant (ca. 40%), with negligible levels of the C_7 conformer.

Some of the previous infrared results were challenged by Maxfield et al.,[159] who found that in $CHCl_3$, the broad NH stretching band below 3400 cm^{-1} in the spectrum of AcGlyNHMe, AcAlaNHMe and AcLeuNHMe has a concentration-dependent intensity. The band is not detectable at 10^{-4} M, and thus must arise from intermolecularly H-bonded species. This band was previously assigned to the hydrogen-bonded NH proton in the C_7 conformer. Maxfield et al.[159] do not argue that conformers in the C_7^{eq} region are absent, but conclude that a strong intramolecular H-bond does not exist in these conformers. They conclude that the infrared spectra support the coexistence of a number of conformers in the blocked amino acids, with considerable flexibility in the molecule.

Information from NMR in the form of coupling constants and NOEs was discussed in the previous subsection in the context of the work of Madison and Kopple.[117] Earlier, Bystrov et al.[153] had interpreted the relatively high coupling constant observed for Ac-D-Ala-D-AlaNHMe in $CHCl_3$ as evidence for a C_7^{ax} conformer in that solvent, and the lower coupling constant in more polar solvents, especially water, as resulting from a shift toward the α_R conformation.

In several studies,[117,121,138,160] potential energy maps have been tested by using the derived conformer populations to calculate the conformationally averaged $^3J_{\alpha N}$. This was first attempted by Ramachandran and Chandrasekaran,[160] who calculated the coupling constant for AcAlaNHMe, with and without an intramolecular H-bond. They obtained values which differed by 0.4 Hz, with the H-bonded form having the smaller J. This is contrary to the usual interpretation of the experimental results,[117] which yield a higher coupling constant for AcAlaNHMe in $CHCl_3$ (7.5 Hz) than in water (6.0 Hz). Madison and Kopple[117] calculated $^3J_{\alpha N}$ values for AcAlaNHMe ranging from 6.2 to 7.4 using various models designed to mimic the effects of hydration. With their standard model, neglecting solvent, they obtained a value of 7.4 Hz, in good agreement with the observed value in $CHCl_3$ (7.5 Hz). Grant et al.[138] obtained values of $^3J_{\alpha N}$ ranging from 6 to 10 Hz, depending on the set of empirical constants used in the Karplus equation and upon the choice of radii[140–142] for calculating the cavity radius in their combination of ab initio and reaction field methods. Their minimum calculated value agrees well with the experimental value of 6.1 Hz.[117] They concluded that the relatively low value of $^3J_{\alpha N}$ in water probably reflects dominant conformers with ϕ near the α_R and P_{II} conformation.

Fermandjian et al.[121] have measured both coupling constants across the NC^α bond, $^3J_{HNC\alpha H}$ ($^3J_{HH}$) and $^3J_{CNC\alpha H}$ ($^3J_{CH}$), for a series of blocked amino acids (Gly,

Ala, Leu, Ile, and tLeu, the latter amino acid being an isomer of Leu in which the side chain is a *tert*-butyl group) in four nonaqueous solvents (dimethylsulfoxide, chloroform, acetonitrile, and dioxane). For blocked Ala, $^3J_{HH}$ ranged from 7.2 Hz in acetonitrile to 8.1 Hz in dioxane, whereas $^3J_{CH}$ ranged from 2.3 Hz in dimethyl-sulfoxide to 2.7 Hz in dioxane. $^3J_{HH}$ increased from an average value of 5.6 Hz for blocked Gly to 9.5 Hz for blocked tLeu. The average $^3J_{CH}$ was smallest for blocked Ala (2.5 Hz) and largest for blocked Gly (3.5 Hz), although the range of values for blocked Gly overlaps that for blocked tLeu. Conformational energy calculations by the SIBFA method[119] correctly reproduced the ordering of both the $^3J_{HH}$ and $^3J_{CH}$ coupling constants, except for predicting a larger value of $^3J_{CH}$ for blocked tLeu than for blocked Gly. The conformational energy maps are dominated by the C_7^{eq} and C_5 conformers for blocked Ala, Leu and Ile, with decreased representation for C_5 in the latter two cases. This leads to an increase in $^3J_{HH}$. In blocked tLeu, only a narrow range of Ramachandran space is accessible, located near $(\phi,\psi) = -140°, 140°$.

V. THEORETICAL STUDIES OF CD IN UNORDERED POLYPEPTIDES AND POLY(PRO)II

Theoretical calculations on the CD of unordered polypeptides have not succeeded in accounting for the observed spectra. Calculations of rotational strengths and transition energies require the geometry to be specified. Two theoretical treatments have considered only dimers[146,147] and have been discussed in Section IV.B., while two others[161,162] have used Monte Carlo techniques to generate peptides of 5–10 residues with probabilities weighted according to conformational energy calcula-tions. In addition, Madison and Schellman[163] calculated the CD of regions of globular proteins which lack regular secondary structure. None of these calcula-tions have succeeded in accounting for the positive $n\pi^*$ band or the strong negative $\pi\pi^*$ band near 200 nm. In the case of the Monte Carlo calculations, this may be attributable to the potential maps which were used, based upon in vacuo potential parameters. It would be of interest to reexamine this problem using results from more recent conformational energy calculations which take hydration into account (Table 3).

Theoretical calculations of the CD of poly(Pro)II helices have been somewhat more successful, and if Tiffany and Krimm[36] are correct, these should be relevant to the interpretation of the CD of unordered polypeptides, at least at low tempera-tures. First, we must consider the assignment of the CD bands observed for the poly(Pro) helix (Fig. 4). The most exhaustive experimental study of the electronic spectrum of poly(Pro)II was that of Mandel and Holzwarth,[164] who determined the absorption and CD spectra of poly(Pro) in isotropic solution and in films in which the molecules were oriented by stroking. With these data, they were able to analyze the CD spectrum in the $\pi\pi^*$ region into the three components expected from exciton theory:[165] gaussian bands polarized parallel and perpendicular to the helix axis, and a non-gaussian band centered at $\lambda_\perp$. The parallel-polarized component has a rotational strength of 0.41 DBM and is located at 209 nm, while the perpendicular-polarized component is at 201 nm with a rotational strength of −0.64 DBM. The non-gaussian component is centered at 201 nm and is negative on the long-wavelength side, and positive on the short-wavelength side. This combination of CD bands can explain the major features of the poly(Pro) CD spectrum. The overall negative nature of the CD in the $\pi\pi^*$ region is due to the greater strength of the perpendicular-polarized component. The negative lobe of the nongaussian band largely cancels the positive parallel-polarized exciton band. Mandel and Holzwarth did not detect a distinct $n\pi^*$ band and attributed the long-wavelength positive CD to the parallel-polarized $\pi\pi^*$ exciton band. However, the long-wavelength positive band was not fit very well by the resultant of the exciton components, and it is likely that there is a small $n\pi^*$ contribution, probably positive.

Several theoretical studies of the CD of the poly(Pro)II helix have been reported.[163,166–172] Treatments which only consider the mixing of the $n\pi^*$ and $\pi\pi^*$ transitions cannot account for the net negative rotational strength observed for poly(Pro)II in the accessible wavelength region. Most calculations predict a positive band in the CD below 190 nm, which results from the positive lobe of the helix band. Rabenold and Rhodes[173] have shown that an analogous problem with calculations of the α-helix CD spectrum is resolved by including high-energy transitions through a polarizability tensor. Mixing with these high-energy transitions selectively reduces the high-energy lobe of the helix band and almost completely eliminates the corresponding short-wavelength feature which theory has consistently predicted but experiment has not detected. That a similar effect is operating in the poly(Pro) helix spectrum is suggested by the fact that those calculations which include the most extensive mixing with high-energy transitions[169–171] show the best agreement with experiment and lack the positive short-wavelength band characteristic of most calculations.

VI. SUMMARY AND CONCLUSIONS

When I reviewed the CD of polypeptides in 1977[2] I concluded that Tiffany and Krimm's model[36] for ionized poly(Glu) and poly(Lys) was not convincingly supported by the existing data. I stated that "... the evidence favors the interpretation that random-coil polypeptides of different constitution and under different conditions can give a variety of CD curves. It is possible that under conditions of high charge density or at low temperatures, some regions of conformational space may be particularly favored and indeed these may be in the poly(Pro)II region. Since local interactions are dominant in determining the CD, this may explain the observation of significant CD in a random coil, the noncoincidence of the absorption and CD maxima, and the strong dependence on temperature and ionic strength. However, this does not imply that significant helical regions occur."

How do more recent developments affect this conclusion? The CD measurements of Drake et al.[6] on ionized poly(Lys) clearly extend into a regime where there is a very strong preference for a narrow range of conformations. At −100°C, where the CD amplitudes seem to be very close to limiting values, it is likely that the range of conformational angles is sufficiently narrow that the polypeptide can indeed be considered helical. At room temperature, the lower amplitudes indicate that while there is still a significant population of residues in the favored conformational minimum, the conformation may either be a frequently interrupted helix, or a long helix with increased deviations from the most favored residue conformation.

The VCD studies of Yasui and Keiderling[7] and of Paterlini et al.[8] provide strong, but not absolutely conclusive, evidence for left-handed helical conformations. However, the short-range nature of vibrational coupling means that a predominance of residue conformations with ϕ, ψ values in a left-handed helical region would be consistent with the results.

There is now good evidence that short stretches of the left-handed three-fold helix occur in globular proteins,[103,104] and that this conformation is dominant in some glycoproteins.[90,91,93,94,99,101] CD and NMR studies[117,148] of blocked amino acids indicate that the P_{II} conformer is likely to be significant in such systems in aqueous solution.

Theoretical studies of blocked amino acids in aqueous solution have been critical in providing a rationale for the widespread preference for the P_{II} conformer at the level of the individual residue. Conformational energy maps based upon in vacuo calculations (Table 2) almost uniformly predict the C_7^{eq} conformer as being the most stable. The map of Brant et al.[110] is exceptional in placing the P_{II} conformer at the global minimum. Even here, though, P_{II} was only slightly more stable than a wide range of conformers in the upper left quadrant of the Ramachandran map. Although disagreeing in detail, the more recent calculations which take solvent into account (Table 3) indicate that hydration markedly stabilizes the P_{II} and α_R conformers with respect to the C_7^{eq} and C_5. These results have helped overcome one of the most serious obstacles to accepting Tiffany and Krimm's hypothesis[35]— why should the P_{II} conformation be prominent in so many systems which lack the steric constraints of poly(Pro) and the electrostatic repulsions of charged poly(Lys) and poly(Glu)?

There is now strong evidence in favor of Tiffany and Krimm's proposal.[36] Unordered polypeptides which exhibit a positive band near 218 nm in the low temperature limit do have a substantial amount of the P_{II} conformation. As the temperature is increased, other regions of the Ramachandran map become more populated, especially the α_R region. This does not mean that these polypeptides should be envisioned as true P_{II} helices, except in the extreme low-temperature limit, such as that approached by Drake et al.[6] at $-100°C$, or when stabilized by carbohydrates as in the antifreeze glycoproteins.[90,91,93,94,99,101] At room temperature, and even near $0°C$, short stretches in the P_{II} conformation will be interspersed with residues having other conformations.

In the case where the $n\pi^*$ CD is negative over the conventional temperature range, conformations in the β and/or α_R region are dominant. This can come about in at least two ways: (1) one or more strong free-energy minima occur in regions other than that of the P_{II} conformation; (2) the minimum in the P_{II} region is much shallower and broader. These two possibilities can, in principle, be distinguished by extending measurements to low temperatures.

ACKNOWLEDGMENTS

This work was supported by an NIH grant, GM22994, and by a Fogarty Senior International Fellowship. I thank Professors Wilfred van Gunsteren and Herman Berendsen of the University of Groningen for their hospitality during the tenure of the Fogarty Fellowship. I am also grateful to Professor R.L. Baldwin for stimulating my interest in this problem and to Professor W.A. Gibbons for a helpful conversation. I also thank Dr. N. Sreerama for a critical reading of the manuscript. Randy DeBey and M.M. Miller assisted in generating two of the figures.

REFERENCES

1. Sears, D. W.; Beychok, S. In *Physical Principles and Techniques of Protein Chemistry;* Leach, S. J., Ed.; Academic Press: New York, 1973; Part C, pp. 445–593.
2. Woody, R. W. *J. Polym. Sci. Macromol. Rev.* **1977**, *12*, 181–320.
3. Baldwin, R. L. *Trends Biochem. Sci. (Ref. Ed.)* **1986**, *11*, 6–9.
4. Wright, P. E.; Dyson, H. J.; Lerner, R. A. *Biochemistry* **1988**, *27*, 7167–7175.
5. Creighton, T. E. *Biophys. Chem.* **1988**, *31*, 155–162.
6. Drake, A. F.; Siligardi, G.; Gibbons, W. A. *Biophys. Chem.* **1988**, *31*, 143–146.
7. Yasui, S. C.; Keiderling, T. A. *J. Am. Chem. Soc.* **1986**, *108*, 5576–5581.
8. Paterlini, M. G.; Freedman, T. B.; Nafie, L. A. *Biopolymers* **1986**, *25*, 1751–1765.
9. Keiderling, T. A. *Nature* **1986**, *322*, 851–852.
10. Kim, P. S.; Baldwin, R. L. *Ann. Rev. Biochem.* **1990**, *59*, 631–660.
11. Holzwarth, G.; Doty, P. *J. Am. Chem. Soc.* **1965**, *87*, 218–228.
12. Schellman, J. A.; Schellman, C. G. In *The Proteins, 2nd ed.*; H. Neurath, Ed., Academic Press: New York, **1964**; Vol. 2, pp. 1—137.
13. Ramachandran, G. N.; Sasisekharan, V.; Ramakrishnan, C. *J. Mol. Biol.* **1963**, *7*, 95–99.
14. Ramachandran, G. N.; Sasisekharan, V. *Adv. Protein Chem.* **1968**, *23*, 283–437.
15. Beychok, S. *Poly-α-Amino Acids*; G. D. Fasman, Ed. Marcel Dekker: New York, 1967; pp. 293–337.
16. Kosen, P. A.; Creighton, T. E.; Blout, E. R. *Biochemistry* **1981**, *20*, 5744–5754.
17. Visser, L.; Blout, E. R. *Biochemistry* **1971**, *10*, 743–752.
18. Huber, R.; Kukla, E.; Bode, W.; Schwager, P.; Bartels, K.; Deisenhofer, J.; Steigemann, W. *J. Mol. Biol.* **1974**, *89*, 73–101.
19. Birktoft, J. J.; Blow, D. M. *J. Mol. Biol.* **1972**, *68*, 187–240.

20. Manning, M. C.; Woody, R. W. *Biochemistry* **1989**, *28*, 8609–8613.
21. Timasheff, S. N.; Susi, H.; Townend, R.; Stevens, L.; Gorbunoff, M. J.; Kumosinski, T. F. In *Conformation of Biopolymers*; G. N. Ramachandran, Ed.; Academic Press: London, 1967; Vol. 1, pp. 173–196.
22. Adler, A. J.; Hoving, R.; Poter, J.; Wells, M.; Fasman, G. D. *J. Am. Chem. Soc.* **1968**, *90*, 4736–4738.
23. Mattice, W. L. *Biopolymers* **1974**, *13*, 169–183.
24. Shoemaker, K. R.; Fairman, R.; York, E. J.; Stewart, J. M; Baldwin, R. L. In *Peptides: Chemistry and Biology, Proc. Tenth Am. Pept. Sympos.*; G.R. Marshall, Ed.; ESCOM: Leiden, **1988**, pp. 15–20.
25. Klee, W. A. *Biochemistry* **1968**, *7*, 2731–2736.
26. Brown, J. E.; Klee, W. A. *Biochemistry* **1971**, *10*, 470–476.
27. Brack, A.; Spach, G. *J. Am. Chem. Soc.* **1981**, *103*, 6319–6323.
28. Dearborn, D. G.; Wetlaufer, D. B. *Biochem. Biophys. Res. Commun.* **1970**, *39*, 314–320.
29. Fasman, G. D.; Hoving, H.; Timasheff, S. N. *Biochemistry* **1970**, *9*, 3316–3324.
30. Cortijo, M.; Panijpan, B.; Gratzer, W. B. *Intl. J. Pept. Protein Res.* **1973**, *5*, 179–186.
31. Privalov, P. L.; Tiktopulo, E. I.; Venyaminov, S. Yu.; Griko, Yu. V.; Makhatadze, G. I.; Khechinashvili, N. N. *J. Mol. Biol.* **1989**, *205*, 737–750.
32. Velluz, L.; Legrand, M. *Angew. Chem. Intl. Ed. Engl.*, **1965**, *4*, 838–845.
33. Carver, J. P.; Schechter, E.; Blout, E. R. *J. Am. Chem. Soc.* **1965**, *88*, 2550–2561.
34. Greenfield, N.; Fasman, G. D. *Biochemistry* **1969**, *8*, 4108–4116.
35. Myer, Y. P. *Macromolecules* **1969**, *2*, 624–628.
36. Tiffany, M. L.; Krimm, S. *Biopolymers* **1968**, *6*, 1379–1382.
37. Tiffany, M. L.; Krimm, S. *Biopolymers* **1969**, *8*, 347–359.
38. Tiffany, M. L.; Krimm, S. *Biopolymers* **1972**, *11*, 2309–2316.
39. Epand, R. M.; Wheeler, G. E.; Moscarello, M. A. *Biopolymers* **1974**, *13*, 359–369.
40. Rao, S. P.; Miller, W. G. *Biopolymers* **1973**, *12*, 835–843.
41. Johnson, W. C., Jr.; Tinoco, I., Jr. *J. Am. Chem. Soc.* **1972**, *94*, 4389–4390.
42. Mattice, W. L.; Lo, J.-T.; Mandelkern, L. *Macromolecules* **1972**, *5*, 729–734.
43. Mattice, W. L.; Lo, J.-T. *Macromolecules* **1972**, *5*, 734–739.
44. Mandelkern, L.; Mattice, W. L. In *Conformation of Biological Molecules and Polymers*; Bergmann, E. D.; Pullman, B., Eds., Israel Academy of Sciences and Humanities: Jerusalem, **1973**, pp. 128–139.
45. Rai, J. H.; Miller, W. G. *Biopolymers* **1973**, *12*, 845–856.
46. Krimm, S.; Mark, J. E. *Proc. Natl. Acad. Sci. USA* **1968**, *60*, 1122–1129.
47. Tiffany, M. L.; Krimm, S. *Biopolymers* **1973**, *12*, 575–587.
48. Tiffany, M. L.; Krimm, S. *Biopolymers* **1968**, *6*, 1767–1770.
49. Nielsen, E. B.; Schellman, J. A. *J. Phys. Chem.* **1967**, *71*, 2297–2304.
50. Krimm, S.; Mark, J. E.; Tiffany, M. L. *Biopolymers* **1969**, *8*, 695–697.
51. Hiltner, W. A.; Hopfinger, A. J.; Walton, A. G. *J. Am. Chem. Soc.* **1972**, *94*, 4324–4327.
52. Swenson, C. A. *Biopolymers* **1971**, *10*, 2591–2596.
53. Torchia, D. A.; Bovey, F. A. *Macromolecules* **1971**, *4*, 246–251.
54. Dorman, D. E.; Torchia, D. A.; Bovey, F. A. *Macromolecules* **1973**, *6*, 80–82.
55. Clark, D. S.; Dechter, J. J.; Mandelkern, L. *Macromolecules* **1979**, *12*, 626–633.
56. Piez, K. A.; Sherman, M. R. *Biochemistry* **1970**, *9*, 4129–4140.
57. Jenness, D. D.; Sprecher, C.; Johnson, W. C., Jr. *Biopolymers* **1976**, *15*, 513–521.
58. Hesselink, F. T.; Ooi, T.; Scheraga, H. A. *Macromolecules* **1973**, *6*, 541–552.
59. Tanford, C.; Kawahara, K.; Lapanje, S. *J. Am. Chem. Soc.* **1967**, *89*, 729–736.
60. Tanford, C. *Adv. Protein Chem.* **1968**, *23*, 121–282.
61. Lapanje, S.; Tanford, C. *J. Am. Chem. Soc.* **1967**, *89*, 5030–5033.
62. Tanford, C. *Adv. Protein Chem.* **1970**, *24*, 1–95.

63. Robinson, D. R.; Jencks, W. P. *J. Am. Chem. Soc.* **1965**, *87*, 2462–2470.

64. Lee, J. C.; Timasheff, S. N. *Biochemistry*, **1974**, *13*, 257–265.

65. Pfeil, W.; Privalov, P. L. *Biophys. Chem.* **1976**, *4*, 41–50.

66. Painter, P. C.; Koenig, J. L. *Biopolymers* **1976**, *15*, 229–240.

67. Chirgadze, Yu. N.; Shestopalov, B. V.; Venyaminov, S. Yu. *Biopolymers* **1973**, *12*, 1337–1351.

68. Painter, P. C.; Coleman, M. M. *Biopolymers* **1978**, *17*, 2475–2484.

69. Smith, M.; Walton, A. G.; Koenig, J. L. *Biopolymers* **1969**, *8*, 173–179.

70. Sengupta, P. K.; Krimm, S. *Biopolymers* **1987**, *26*, S99–S107.

71. Sengupta, P. K.; Krimm, S. *Biopolymers* **1985**, *24*, 1479–1491.

72. Keith, H. D.; Giannoni, G.; Padden, F. J., Jr. *Biopolymers* **1969**, *7*, 775–792.

73. Keith, H. D. *Biopolymers* **1971**, *10*, 1099–1101.

74. Koenig, J. L.; Frushour, B. *Biopolymers* **1972**, *11*, 1871–1892.

75. Sugawara, Y.; Harada, I.; Matsuura, H.; Shimanouchi, T. *Biopolymers* **1978**, *17*, 1405–1421.

76. Sengupta, P. K.; Krimm, S.; Hsu, S. L. *Biopolymers* **1984**, *23*, 1565–1594.

77. Yu, T.-J.; Lippert, J. L.; Peticolas, W. L. *Biopolymers* **1973**, *12*, 2161–2176.

78. Yasui, S. C.; Keiderling, T. A. *Biopolymers* **1986**, *25*, 5–15.

79. Lal, B. B.; Nafie, L. A. *Biopolymers* **1982**, *21*, 2161–2183.

80. Sen, A. C.; Keiderling, T. A. *Biopolymers* **1984**, *23*, 1519–1532.

81. Kurz, J.; Berger, A.; Katchalski, E. *Nature* **1956**, *178*, 1066–1067.

82. Woolley, S.-Y. C.; Holzwarth, G. *Biochemistry* **1970**, *9*, 3604–3608.

83. Dukor, R. K.; Keiderling, T A. *Proc. 20th Eur. Pept. Symp.*; Bayer, E.; Jung, G., Eds.; W. de Gruyter: Berlin, 1989; pp. 519–521.

84. Kobrinskaya, R.; Yasui, S. C.; Keiderling, T. A. In *Peptides: Chemistry and Biology, Proc. Tenth Am. Pept. Sympos.*; Marshall, G. R., Ed.; ESCOM: Leiden, 1988, pp. 65–67.

85. Dukor, R. K.; Keiderling, T. A.; Gut, V. *Int. J. Pept. Protein Res.* **1991**, *38*, 198–203.

86. Okabayashi, H.; Isemura, T.; Sakakibara, S. *Biopolymers* **1968**, *6*, 323–330.

87. Rippon, W. B.; Walton, A. G. *Biopolymers* **1971**, *10*, 1207–1212.

88. Rippon, W. B.; Chen, H. H.; Walton, A. G. *J. Mol. Biol.* **1973**, *75*, 369–375.

89. Bush, C. A.; Feeney, R. E.; Osuga, D. T.; Ralapati, S.; Yeh, Y. *Int. J. Pept. Protein Res.* **1981**, *17*, 125–129.

90. Bush, C. A.; Ralapati, S.; Matson, G. M.; Yamasaki, R. B.; Osuga, D. T.; Yeh, Y.; Feeney, R. E. *Arch. Biochem. Biophys.* **1984**, *232*, 624–631.

91. Ahmed, A. I.; Feeney, R. E.; Osuga, D. T.; Yeh, Y. *J. Biol. Chem.* **1975**, *250*, 3344–3347.

92. Berman, E.; Allerhand, A.; DeVries, A. L. *J. Biol. Chem.* **1980**, *255*, 4407–4410.

93. Bush, C. A.; Feeney, R. E. *Int. J. Pept. Protein Res.* **1986**, *28*, 386–397.

94. Rao, B. N. N.; Bush, C. A. *Biopolymers* **1987**, *26*, 1227–1244.

95. Filira, F.; Biondi, L.; Scolaro, B.; Foffani, M. T.; Mammi, S.; Peggion, E.; Rocchi, R. *Int. J. Biol. Macromol.* **1990**, *12*, 41–49.

96. Avanova, A. Ya. *Mol. Biol.* **1990**, *24*, 581–597.

97. Avanov, A. Ya.; Lipkind, G. M.; Kochetkov, N. K. *Bioorg. Khim.* **1982**, *8*, 616–620.

98. Homans, S. W.; de Vries, A. L.; Parker, S. B. *FEBS Lett.* **1985**, *183*, 133–137.

99. van Holst, G.-J.; Varner, J. E. *Plant Physiol.* **1984**, *74*, 247–251.

100. Matsumoto, I.; Jimbo, A.; Mizuno, Y.; Seno, N.; Jeanloz, R. W. *J. Biol. Chem.* **1983**, *258*, 2886–2891.

101. van Holst, G.-J.; Martin, S. R.; Allen, A. K.; Ashford, D.; Desai, N. N.; Neuberger, A. *Biochem. J.* **1986**, *233*, 731–736.

102. Allen, A. K.; Desai, N. N.; Neuberger, A.; Creeth, J. M. *Biochem. J.* **1978**, *171*, 665–674.

103. Adzhubei, A. A.; Eisenmenger, F.; Tumanyan, V. G.; Zinke, M.; Brodzinski, S.; Esipova, N. G. *Biochem. Biophys. Res. Commun.* **1987**, *146*, 934–938.

104. Adzhubei, A. A.; Eisenmenger, F.; Tumanyan, V. G.; Zinke, M.; Brodzinski, S.; Esipova, N. G. *J. Biomol. Struct. Dyn.* **1987**, *5*, 689–704.

105. IUPAC-IUB Commission on Biochemical Nomenclature *Biochemistry* **1970**, *9*, 3471–3479.

106. Bernstein, F. C.; Koetzle, T. F.; Williams, G. J. B.; Meyer, E. F., Jr.; Brice, M. D.; Rodgers, J. R.; Kennard, O.; Shimanouchi, T.; Tasumi, M. *J. Mol. Biol.* **1977**, *112*, 535–542.

107. Pauling, L.; Corey, R. B. *Proc. Natl. Acad. Sci. USA* **1951**, *37*, 729–740.

108. Makarov, A. A.; Esipova, N. G.; Lobachov, V. M.; Grishovsky, B. A.; Pankov, Yu. A. *Biopolymers* **1984**, *23*, 5–22.

109. Scott, R. A.; Scheraga, H. A. *J. Chem. Phys.* **1966**, *45*, 2091–2101.

110. Brant, D. A.; Miller, W. G.; Flory, P. J. *J. Mol. Biol.* **1967**, *23*, 47–65.

111. Hoffman, R.; Imamura, I. A. *Biopolymers* **1969**, *7*, 207–213.

112. Maigret, B.; Perahia, D.; Pullman, B. *J. Theor. Biol.* **1970**, *29*, 275–291.

113. Momany, F. A.; McGuire, R. F.; Yan, J. F.; Scheraga, H. A. *J. Phys. Chem.* **1971**, *75*, 2286–2297.

114. Lewis, P. N.; Momany, F. A.; Scheraga, H. A. *Isr. J. Chem.* **1973**, *11*, 121–152.

115. Zimmerman, S. S.; Pottle, M. S.; Némethy, G.; Scheraga, H. A. *Macromolecules* **1977**, *10*, 1–9.

116. Hillier, I. H.; Robson, B. *J. Theor. Biol.* **1979**, *76*, 83–98.

117. Madison, V.; Kopple, K. D. *J. Am. Chem. Soc.* **1980**, *102*, 4855–4863.

118. Rossky, P. J.; Karplus, M.; Rahman, A. *Biopolymers* **1979**, *18*, 825–854.

119. Gresh, N.; Pullman, A.; Claverie, P. *Theor. Chim. Acta* **1985**, *67*, 11–32.

120. Roterman, I. K.; Lambert, M. H.; Gibson, K. D.; Scheraga, H. A. *J. Biomol. Struct. Dynam.* **1989**, *7*, 421–453.

121. Fermandjian, S.; Sakarellos, C.; Aumelas, A.; Toma, F.; Gresh, N. *Intl. J. Pept. Protein Res.* **1990**, *35*, 473–480.

122. Hoffmann, R. *J. Chem. Phys.* **1963**, *39*, 1397–1412.

123. Pople, J. A.; Segal, G. A. *J. Chem. Phys.* **1966**, *44*, 3289–3296.

124. Diner, S.; Malrieu, J. P.; Claverie, P. *Theor. Chim. Acta* **1969**, *13*, 1–17.

125. Momany, F. A.; McGuire, R. F.; Burgess, A. W.; Scheraga, H. A. *J. Phys. Chem.* **1975**, *79*, 2361–2381.

126. Brooks, B. R.; Bruccoleri, R. E.; Olafson, B. D.; States, D. J.; Swaminathan, S.; Karplus, M. *J. Comp. Chem.* **1983**, *4*, 187–217.

127. Weiner, P. K.; Kollman, P. A. *J. Comp. Chem.* **1981**, *2*, 287–303.

128. Némethy, G.; Pottle, M. S.; Scheraga, H. A. *J. Phys. Chem.* **1983**, *87*, 1883–1887.

129. Pullman, B.; Maigret, B.; Perahia, D. *Theor. Chim. Acta* **1970**, *18*, 44–56.

130. Hodes, Z. I.; Némethy, G.; Scheraga, H. A. *Biopolymers* **1979**, *18*, 1565–1610.

131. Hodes, Z. I.; Némethy, G.; Scheraga, H. A. *Biopolymers* **1979**, *18*, 1611–1634.

132. Mezei, M.; Mehrotra, P. K.; Beveridge, D. L. *J. Am. Chem. Soc.* **1985**, *107*, 2239–2245.

133. Pettitt, B. M.; Karplus, M. *Chem. Phys. Lett.* **1985**, *121*, 194–201.

134. Lau, W. F.; Pettitt, B. M. *Biopolymers* **1987**, *26*, 1817–1831.

135. Anderson, A. G.; Hermans, J. *Proteins* **1988**, *3*, 262–265.

136. Renugopalakrishnan, V.; Nir, S.; Rein, R. In *Environmental Effects on Molecular Structure and Properties*, B. Pullman, Ed.; D. Reidel: Dordrecht, 1976; pp. 109–133.

137. Pettitt, B. M.; Karplus, M.; Rossky, P. J. *J. Phys. Chem.* **1986**, *90*, 6335–6345.

138. Grant, J. A.; Williams, R. L.; Scheraga, H. A. *Biopolymers* **1990**, *30*, 929–949.

139. Forsythe, K. H.; Hopfinger, A. J. *Macromolecules* **1973**, *6*, 423–437.

140. Rashin, A.; Namboodiri, K. *J. Phys. Chem.* **1987**, *91*, 6003–6012.

141. Aguilar, M. A.; Martin, M. A.; Tolosa, S.; Olivares Del Valle, F. J. *J. Mol. Struct. (Theochem)* **1988**, *166*, 313–318.

142. Aguilar, M. A.; Olivares Del Valle, F. J. *Chem. Phys.* **1989**, *129*, 439–450.

143. Bayley, P. M.; Nielsen, E. B.; Schellman, J. A. *J. Phys. Chem.* **1969**, *73*, 228–243.

144. Nielsen, E. B.; Schellman, J. A. *Biopolymers* **1971**, *10*, 1559–1581.

145. Moscowitz, A. In *Optical Rotatory Dispersion;* Djerassi, C., Ed.; McGraw-Hill: New York, 1960; pp. 150–177.

146. Tonelli, A. E. *Macromolecules* **1969**, *2*, 635–637.

147. Ronish, E. W.; Krimm, S. *Biopolymers* **1972**, *11*, 1919–1928.

148. Dungan, J. M., III; Hooker, T. M., Jr. *Macromolecules* **1981**, *14*, 1812–1822.

149. Siemion, I. Z.; Wieland, T.; Pook, K. H. *Angew. Chem. Int. Ed. Engl.* **1975**, *14*, 702–703.

150. Mizushima, S.; Shimanouchi, T.; Tsuboi, M.; Arakawa, T. *J. Am. Chem. Soc.* **1957**, *79*, 5357–5361.

151. Tsuboi, M.; Shimanouchi, T.; Mizushima, S. *J. Am. Chem. Soc.* **1959**, *81*, 1406–1411.

152. Portnova, S. L.; Bystrov, V. F.; Tsetlin, V. I.; Ivanov, V. I.; Ovchinnikov, Yu. A. *Zh. Obshch. Khim.* **1968**, *38*, 428–439.

153. Bystrov, V. F.; Portnova, S. L.; Tsetlin, V. I.; Ivanov, V. T.; Ovchinnikov, Yu. A. *Tetrahedron* **1969**, *25*, 493–515.

154. Marraud, M.; Néel, J.; Avignon, M.; Huong, P. V. *J. Chim. Phys. Phys.-Chim. Biol.* **1970**, *67*, 959–964.

155. Avignon, M.; Huong, P. V.; Lascombe, J.; Marraud, M.; Néel, J. *Biopolymers* **1969**, *8*, 69–89.

156. Avignon, M.; Huong, P. V. *Biopolymers* **1970**, *9*, 427–432.

157. Burgess, A. W.; Scheraga, H. A. *Biopolymers* **1973**, *12*, 2177–2183.

158. Cung, M. T.; Marraud, M.; Néel, J. In *Conformation of Biological Molecules and Polymers*; Bergmann, E. D.; Pullman, B., Eds., Academic Press: New York, 1973, pp. 69–86.

159. Maxfield, F. R.; Leach, S. J.; Stimson, E. R.; Powers, S. P.; Scheraga, H. A. *Biopolymers* **1979**, *18*, 2507–2521.

160. Ramachandran, G. N.; Chandrasekaran, R. *Biopolymers* **1971**, *10*, 935–939.

161. Aebersold, D.; Pysh, E. S. *J. Chem. Phys.* **1970**, *53*, 2156–2163.

162. Zubkov, V. A.; Birshtein, T. M.; Milevskaya, I. S.; Volkenstein, M. V. *Biopolymers* **1971**, *10*, 2051–2061.

163. Madison, V.; Schellman, J. *Biopolymers* **1972**, *11*, 1041–1076.

164. Mandel, R.; Holzwarth, G. *Biopolymers* **1973**, *12*, 655–674.

165. Tinovo, I., Jr. *J. Am. Chem. Soc.* **1964**, *86*, 297–298.

166. Pysh, E. S. *J. Mol. Biol.* **1967**, *23*, 587–599.

167. Pysh, E. S. *J. Chem. Phys.* **1970**, *52*, 4723–4733.

168. Tterlikkis, L.; Loxsom, F. M.; Rhodes, W. *Biopolymers* **1973**, *12*, 675–684.

169. Pysh, E. S. *Biopolymers* **1974**, *13*, 1563–1571.

170. Ronish, E. W.; Krimm, S. *Biopolymers* **1974**, *13*, 1635–1651.

171. Applequist, J. *Biopolymers* **1981**, *20*, 2311–2322.

172. Manning, M. C.; Woody, R. W. *Biopolymers* **1991**, *31*, 569–586.

173. Rabenold, D. A.; Rhodes, W. *J. Phys. Chem.* **1986**, *90*, 2560–2566.

174. Edsall, J. T.; Flory, P. J.; Kendrew, J. C.; Liquori, A. M.; Némethy, G.; Ramanchandran, G. N.; Scheraga, H. A. *J. Biol. Chem.* **1966**, *241*, 1004–1008.

LUMINESCENCE STUDIES WITH HORSE LIVER ALCOHOL DEHYDROGENASE:
INFORMATION ON THE STRUCTURE, DYNAMICS, TRANSITIONS AND INTERACTIONS OF THIS ENZYME

Maurice R. Eftink

Advances in Biophysical Chemistry, Volume 2, pages 81–114
Copyright © 1992 by JAI Press Inc.
All rights of reproduction in any form reserved.
ISBN: 1-55938-396-8

I. INTRODUCTION

Various luminescence techniques are well established for the determination of information about the solution structure and dynamics, as well as thermodynamic and kinetic properties, of proteins and other organized biomolecules. Fluorescence and phosphorescence methods are very versatile and convenient, and they require only small, dilute samples. Depending on the type of measurement, a researcher can obtain information about the solvent (or quencher) accessibility of reporter groups, the existence of different conformational states, the distance between reporter groups, and the rotational freedom of groups. Additionally, fluorescence methods can be used to monitor the kinetics and thermodynamics of various transitions (i.e., ligand binding, denaturation, etc.). These features make fluorescence techniques among the more popular in a modern biophysical laboratory. Numerous studies, with proteins having a single fluorescent reporter group, have revealed the above types of information.

A disadvantage of fluorescence measurements in solution phase is that spectra are often very poorly resolved. Emission and excitation bands are usually broad and have little or no vibronic structure. If there is more than one emitting center in a protein (i.e., a multi-tryptophan protein), then the overlap of various contributions makes interpretations quite difficult. The dissection of various components can sometimes be achieved. When this dissection is possible it can usually be attributed to a) the sensitivity of various luminescence properties to the environment of a lumiphore, and b) the multi-dimensional nature of luminescence. The fluorescence properties of lumiphores, such as the indole side chain of tryptophan, are much more sensitive to solvent polarity, quenchers, temperature, etc., than are its absorbance properties. Consequently there may be a difference in Stokes shifts, fluorescence decay times, and so on, for lumiphores in different environments. The environmental sensitivity, together with the ability to make measurements along multiple dimensions axes (i.e., wavelength, decay time, polarizer angle, quencher concentration) enable some dissection of components to be achieved.

So the versatility, convenience, sensitivity, and the usefulness of the information provided by luminescence techniques are countered by the low resolution of the techniques. For proteins having multiple luminescing probes (i.e., tryptophan residues), the resolution problem can be overwhelming and only the most qualitative interpretations should be made in most cases. Some informative studies have been possible with proteins having two to three reporter groups. Most notable among these is the protein alcohol dehydrogenase (LADH)[1] from horse liver.

LADH is a homo-dimeric protein, having 40 kDa subunits. Each subunit has two types of tryptophan residues, Trp-314, which is buried at the intersubunit interface, and Trp-15, which is a surface residue. The different environment of these two residues results in there being quite distinct fluorescence and phosphorescence properties for the two residues.

Since the structure of LADH and its complexes with coenzymes is known to high resolution, this protein often has served as a test system for the development and interpretation of luminescence methods, before their use with other, less known protein systems. In particular, LADH serves as an example of a multi-tryptophan containing protein for the development of methods to resolve luminescence contributions from individual centers. Thus this protein holds a special place in the field of biophysical luminescence. Our research group is one of several that have studied LADH in this manner.

This article is not a review of luminescence techniques. I suggest the following references for such background information.[2–8] Instead, my attempt is to use the example of the protein LADH to illustrate the broad scope of information that can be obtained with these methods on a particular protein system. I will also try to point out the limitations and ambiguities of these techniques.

II. CRYSTAL STRUCTURE OF LADH

The crystal structure of LADH and several of its derivatives and binary complexes have been reported.[9–13] Shown in Figure 1 is a chain structure of LADH with the trp residues and coenzyme, NADH,[13] highlighted. Each subunit of LADH can be considered to be comprised of two domains, a coenzyme-binding domain, which also includes the contact region between the subunits, and a catalytic domain, which includes the binding site for the alcohol substrates and includes the sites for two zinc ions. One of the zinc ions is at the active site, which lies within the cleft region between the two domains. This zinc is essential for activity. The second zinc ion is in the larger, catalytic domain and is believed to serve a structural role.

In the absence of ligands LADH crystallizes in an orthorhombic space group and its structure has an open active site. Complexes with strongly bound coenzymes form triclinic crystals and the structures show that the binding of coenzyme induces

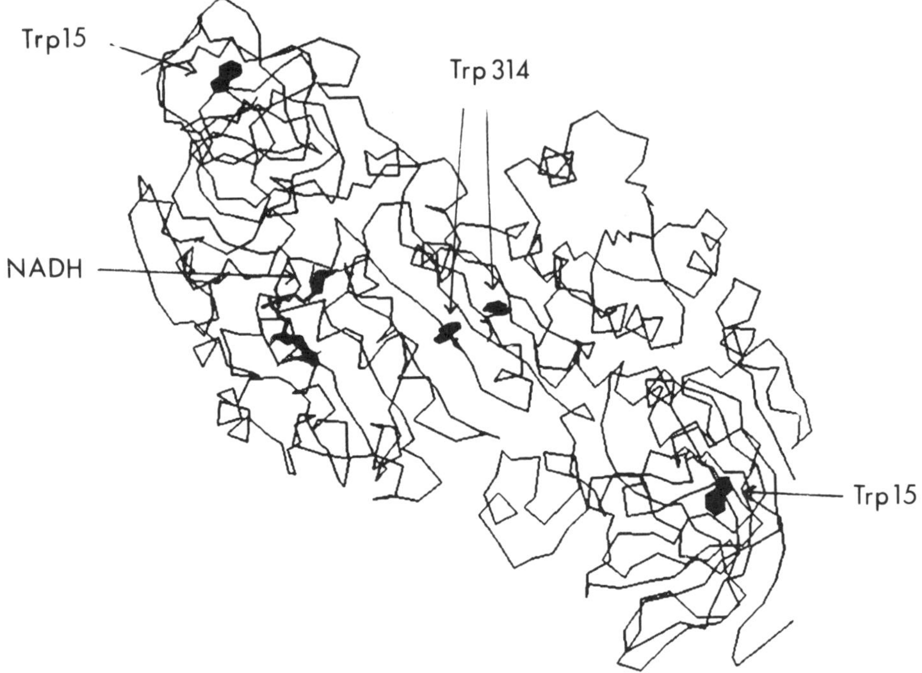

Figure 1. Backbone structure of the complex of LADH with NADH and DMSO. Highlighted are the coenzymes and the two classes of trp residues. The two Trp-314 residues are near the center of the structure and the two Trp-15 are at either end. Only a sinde NADH is shown for clarity.

closing of the active site. This conformational rearrangement from an open to a closed form involves a rotation of the two domains by about 10° with respect to a central axis.

Trp-15 is located at the outer edge of the catalytic domain. Its indole side chain is accessible to solvent. It has been suggested that the indole nitrogen of Trp-15 forms a hydrogen bond with Glu-24 and that this interaction locks the indole ring in place.

Trp-314 is in the coenzyme domain and is part of the subunit-subunit interface region. This apolar core region is formed by β-sheet structures from each subunit. The Trp-314 residues are tightly embedded between these β-structures; no potential interactions between the indole ring and water or polar groups are apparent. The indole ring is in contact with Val-290, Leu-301, and Met-303 of its own subunit and Met-303′ and Leu-308′ of the opposite subunit. In fact, the two Trp-314 residues are very close to one another. The center-to-center distance of their indole rings is only about 6–7Å. The peptide NH and C=O groups of Trp-314 of one subunit form hydrogen bonds with the complementary groups of the other Trp-314.

The coenzyme binds in an extended manner into the cleft between the two domains, with the adenosine portion being inserted into a pocket in the coenzyme domain. The pyrophosphate group is positioned near oppositely charged arginine and lysine residues, and the nicotinamide ring is bound near the bottom of the major cleft, with hydrogen bonds being formed between its nicotinamide group and polar peptide backbone groups of the protein. The alcohol binding site lies between C4 of the nicotinamide ring and the catalytic zinc ion.

III. INTRINSIC AND EXTRINSIC LUMINESCENCE PROBES FOR LADH

In this section a brief description of the intrinsic and extrinsic lumiphores that are available in this protein and its complexes is given. The following sections will then deal with the application of various luminescence techniques to use these lumiphores as probes.

A. Tryptophan Residues

The separate contributions of Trp-314 and -15 to the fluorescence of LADH, suggested by earlier denaturation[14] and low temperature phosphorescence[15] experiments, was first revealed by the iodide quenching studies of Laws and Shore[16] and Abdallah et al.[17] As this quencher is added, there is a blue shift in the emission of the protein. A blue component ($\lambda_{max} \approx 320$ nm, quantum yield = 0.38), comprising about 60% of the total fluorescence of the protein (with excitation at 295 nm), is found to be completely inaccessible to iodide. A red component ($\lambda_{max} \approx 340$ nm, quantum yield = 0.19) is quenched by iodide with Stern–Volmer quenching constant of 5–7 M^{-1}. Subsequent quenching studies with acrylamide showed a similar resolution of inaccessible and accessible components.[18] (With oxygen as quencher there is only a slightly preferred quenching of the red component, as expected with this smaller quenching probe.[19,20] Inspection of the crystal structure of LADH enabled an assignment of the red and blue components to be the buried Trp-314 and surface Trp-15 residues. Additional support for this assignment comes from studies of the binding of coenzymes, which are discussed below. Not only is there a difference in the emission λ_{max} of the two trps, but various luminescence studies show that Trp-314 can be preferentially excited at longer excitation wavelengths.[15,16,17] In Figure 2 are shown the excitation and emission spectra for these two classes of trp residues, as resolved by iodide quenching.

The binding of NADH (and, to a lesser extent, NAD$^+$) to LADH also causes quenching of the tryptophan fluorescence, but in this case there is a red shift on quenching, indicating a preferential quenching of the blue emitting trp residue.[16]

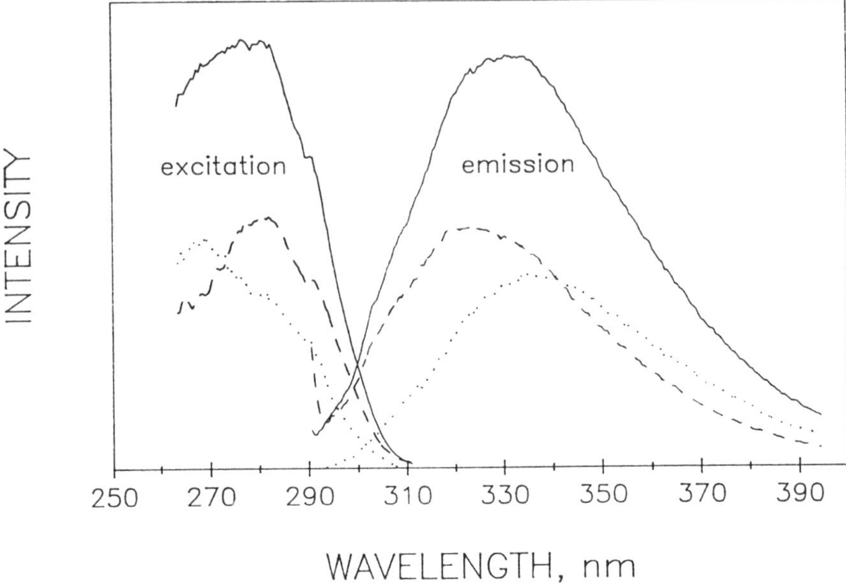

Figure 2. Excitation and emission spectra for LADH (—) and for the Trp-314 (--) and Trp-15 (....) components. The spectra for Trp-314 were obtained from measurements in the presence of 0.75 M KI, to quench ~80% of the contribution from Trp-15. The spectra for Trp-15 were obtained by difference. Conditions, pH 7.2, 0.03 M sodium phosphate, 20 °C, 3 nm slits. Below 265 nm the absorbance by iodide becomes prohibitively large.

Since Trp-314 is nearer to the LADH binding site, this observation is consistent with the above assignment of the blue and red components to Trp-314 and -15, respectively. In Figure 3 are shown relative fluorescence spectra for LADH and its ternary complexes with NAD^+ and pyrazole, and with NAD^+ and trifluoroethanol (TFE). Likewise, in Figure 4 are show the fluorescence spectra of LADH and its complex with NADH and isobutyramide (IB). In Table 1, is summarized a number of luminescence properties for the two trp residues of LADH and its ternary complexes.[21]

In addition to solute and NAD^+ quenching, the fluorescence of the two tryptophans has also been characterized by time-resolved measurements. Using the time correlated single photon counting method, Ross et al.[22] found a bi-exponential decay for LADH, with $\tau_1 = 3.8$ ns and $\tau_2 = 7.2$ ns (at 10°C). The longer lived component was found to have the redder decay associated spectrum and to be the component that is dynamically quenched by iodide. Thus, time-resolved data are consistent with the assignment of Trp-15 to be the longer lived, redder, quencher accessible component, and Trp-314 to be the shorter lived, bluer, inaccessible

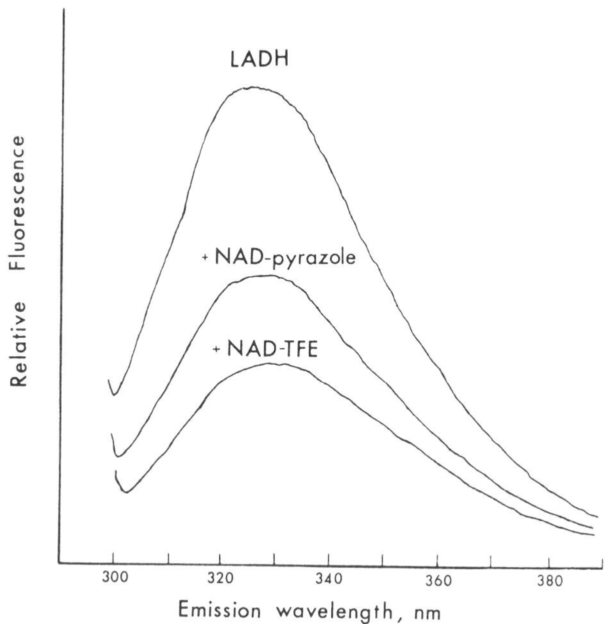

Figure 3. Fluorescence emission spectra of LADH, the LADH NAD⁺ pyrazole ternary complex, and the LADH·NAD⁺·TFE ternary complex. Excitation at 295 nm, 5 nm slits, 20 °C, pH 7.2, 0.03 M sodium phosphate buffer.

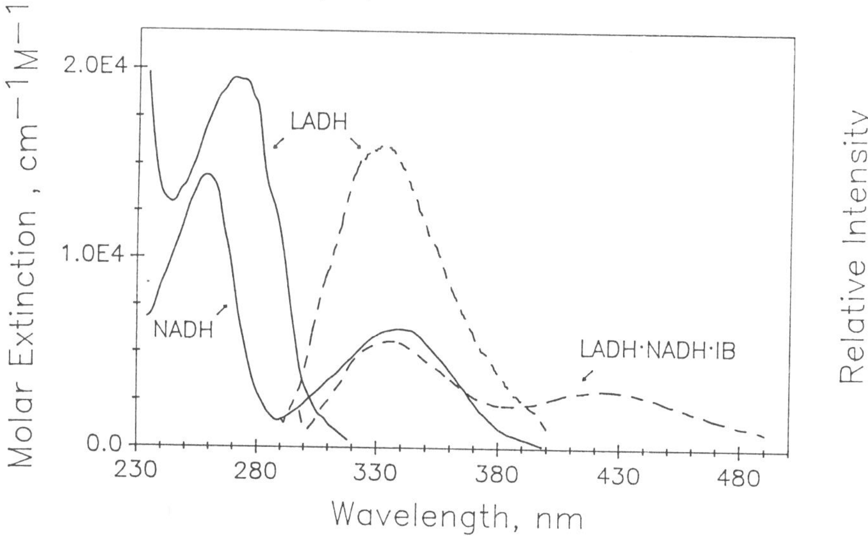

Figure 4. Fluorescence spectra of LADH (--) and the LADH·NADH·IB ternary complex (--) (right ordinant). Excitation at 290 nm, 5 nm slits, pH 7.2, 20 °C. Also shown are the absorption spectrum of LADH (—) and free NADH (—) (left ordinant, given as molar extinction coefficient per site).

Table 1. Selected Luminescence Properties of Trp-15 and Trp-314 of Liver Alcohol Dehydrogenase, Its Complexes and Altered Forms[a]

	Trp-15	*Trp-314*
LADH (apoenzyme)		
fluorescence λ_{max} (nm)	335[18], 337[22], 340[17,45]	320[17], 323[18], 324[22,45]
quantum yield, Φ[a]	0.19[17]	0.37[17]
τ^s (lifetime)(ns)[b]	6.7[23,45], 6.9[19], 7.1[102], 7.2[22], 7.3[24], 7.45[102]	3.4[45], 3.6[19,24], 3.75[102], 3.8[22,23], 3.95[102]
r_o^s (steady state anisotropy)	0.21[24]	0.265[24]
ϕ^s (rotational correlation time)(ns)	36[22]	36[22]
k_q^s by iodide ($\times 10^{-9} M^{-1}s^{-1}$)	0.7[45], 0.76[22], 1.7[18]	~0[18,45]
k_q^s by acrylamide ($\times 10^{-9} M^{-1}s^{-1}$)	1.1, 1.2[18]	~0.01[18]
k_q^s by oxygen ($\times 10^{-9} M^{-1}s^{-1}$)	3.5[19]	0.5[20], 1.0[19]
phosphorescence 0–0 band (nm)	405, 407	410.5[29], 412[15]
τ^T (lifetime)(s)	—	0.30[32], 0.42[28]
k_q^T by acrylamide ($\times 10^{-9} M^{-1}s^{-1}$)	—	0.000002[50]
k_q^T by oxygen ($\times 10^{-9} M^{-1}s^{-1}$)	—	0.6[44], 0.14[40], 0.15[51], 0.036[48]
LADH·NAD$^+$·TFE Complex[c]		
fluorescence λ_{max} (nm)	343[45]	321[45]
τ^s (ns)	6.4[45], 6.98[24]	1.1[45], 1.43[24]
k_q^s by iodide ($\times 10^{-9} M^{-1}s^{-1}$)	1.4[45]	~0[45]
k_q^s by acrylamide ($\times 10^{-9} M^{-1}s^{-1}$)	1.2[24], 1.1[18]	~0[18,24]
r_o^s	0.235[24]	0.290[24]
room temp phosphorescence quenched		
LADH·NAD$^+$·pyrazole Complex[d]		
fluorescence λ_{max} (nm)	~340[23]	~325[23]
τ^s (ns)	7.1[24], 7.0[23], 7.2[22]	2.1[24], 2.4[22,23]
k_q^s by acrylamide ($\times 10^{-9} M^{-1}s^{-1}$)	1.3[24]	~0[24]
r_o^s	0.235[24]	0.285[24]
τ^T (phos. lifetime)(s)	—	0.693[28]
LADH·NADH·IB Complex[e]		
τ^s (ns)	4.5[26]	0.68[26]
τ^T (phos. lifetime)(s)	—	0.635[28]
Zinc Deficient LADH		
τ^s (ns)	8.03[24]	3.3[24]
Acidic Form[f]		
τ^s (ns)	5.3[23]	2.4[23]
Alkaline Form		
τ^s (ns)	7.0[24]	2.2[24]

[a]All values at 20–25 °C and neutral pH, except for the 0–0 phosphorescence peaks, which are at 77 °K and the study by Ross et al.[22] at 10 °C. Literature references for values are indicated as subscripts.

[b]Demmer et al.[25] report a tri-exponential decay fit with lifetimes of 6.82 ns and 2.1 ns assigned to Trp-15 and a lifetime of 4.2 ns assigned to Trp-314.

[c]The LADH·NAD$^+$·TFE ternary complex has 42%,[77] 49%[18] to 53%[28] remaining fluorescence.

[d]The LADH·NAD$^+$·pyrazole ternary complex has 60%,[77] 62%[23] to 67%[28] remaining fluorescence.

[e]The LADH·NADH·IB complex has 24%[28] to 33%[54] remaining fluorescence.

[f]The acid denatured in form of LADH has 42%[23] remaining fluorescence. Some of the variance in % remaining fluorescence may be attributed to differences in excitation wavelength.

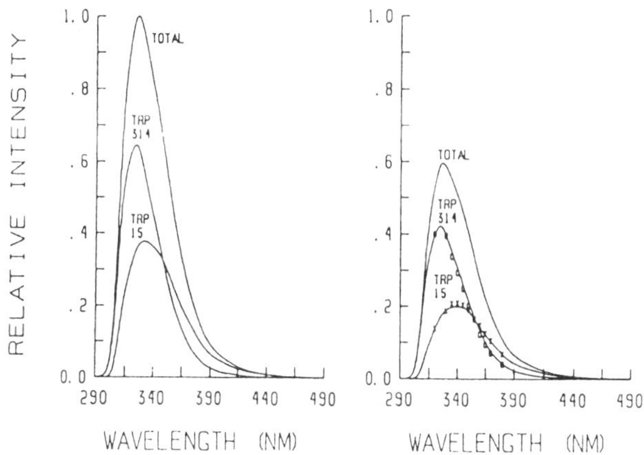

Figure 5. Steady state and decay associated spectra (DAS) for LADH (left panel) and the LADH-NAD⁺·pyrazole ternary complex (right panel). The upper spectra are the total steady-state emission. The lower two curves are the DAS corresponding to fluorescence decay times assigned to Trp-314 and Trp-15. The figure was reproduced from Knutson et al.[23] with permission from the authors and the American Chemical Society.

component. In Figure 5 is shown the decay associated spectra for Trp-15 and -314 in LADH and in a ternary complex of LADH with NAD⁺ and pyrazole. Multifrequency phase/modulation are consistent with this result and assignment. Shown in Figure 6 are phase/modulation data for LADH at different emission wavelengths. In additional time-resolved studies with LADH, Demmer et al.[25] analyzed fluorescence decay data, as a function of iodide concentration, to reveal that the decay law is best described as a triple exponential. A ~4 ns, unquenchable component was still attributed to Trp-314, but Trp-15 was described as being a biexponential with lifetimes (iodide quenchable) of ~7 ns (major component) and ~2 ns (minor component). These researchers proposed that two different conformations of Trp-15 exist with different interactions with nearby Glu-24. The existence of a tri-exponential decay has not been confirmed by other groups, but the ~2 ns component is close to the ~4 ns component and the amplitude of the ~2 ns component is small (about 10%). For these reasons it will be difficult to detect the short-lived component in single time-resolved fluorescence measurements. Only by making measurements as a function of quencher concentration, as done very thoroughly by Demmer and coworkers, will this third component be resolvable.

The binding of NAD⁺ is found to dynamically quench the Trp-314 component from ~3.5 ns in the free enzyme to 2.1 ns in the NAD⁺ pyrazole ternary complex, in agreement with the above steady-state results. The lifetime of Trp-15 is not significantly changed in this complex, but the emission of this residue appears to be slightly red shifted and to perhaps be statically quenched.[23]

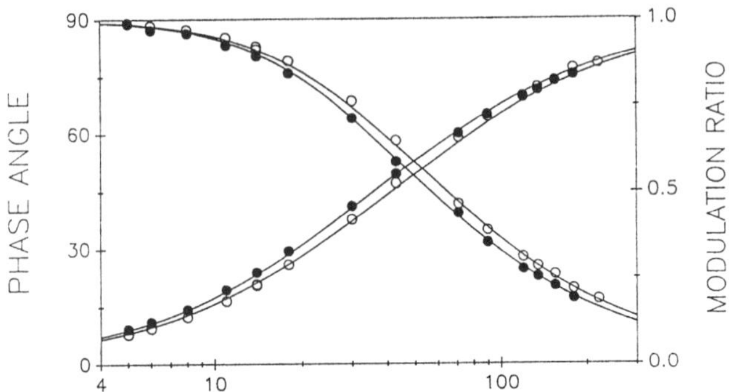

Figure 6. Phase-modulation fluorescence lifetime data for LADH at emission wavelengths of 320 nm (O) and 380 nm (●). Data were also obtained at 340 nm and 360 nm; these data points lie between the points at 320 and 380 nm and are not show for clarity. Conditions, pH 7.2, 0.03 M sodium phosphate, 20 °C, excitation at 300 nm with "magic angle" polarization. The solid lines are global, bi-exponential fits with $\tau_1 = 2.8$ ns and $\tau_2 = 6.3$ ns.

Both Trp-314 and -15 are dynamically quenched by the binding of NADH.[26] The lifetime of Trp-314 drops from 3.5 to 1.5 ns and that of Trp-15 drops from 7 ns to 5 ns. The greater quenching of Trp-314 is consistent with the shorter distance of this residue to the bound dihydronicotinamide group (see following section). Figure 7 shows time-resolved data for LADH and its complex with NADH and isobutyramide.

The separate luminescence of the two trps of LADH was also clearly shown in phosphorescence studies, as first demonstrated by Purkey and Galley.[15] In a low temperature glass (77 °K), two sharp phosphorescence peaks, at 405 nm and 410 nm, were observed for LADH. These peaks are characteristic of the 0–0 vibronic $T_1 \rightarrow S_0$ transition of tryptophan. As temperature is increased the 405 nm peak disappears, as shown in Figure 8, until at room temperature only the 410 nm phosphorescence peak remains. The 410 nm component is attributed to the deeply buried Trp-314, which is protected from quenching by oxygen and other impurities.[27] The triplet state of Trp-314 can be selectively populated at low temperature by excitation at the red edge (305 nm).[15] Additional studies have shown that the binding of NAD^+ and NADH quenches the phosphorescence of Trp-314.[28,29]

B. Bound NADH and Other Probes

In addition to the intrinsic luminescence of the tryptophan residues, the ligand NADH also fluoresces. In fact, there is approximately a 13-fold enhancement (at 410 nm) in the fluorescence of NADH upon binding to LADH in an isobutyramide

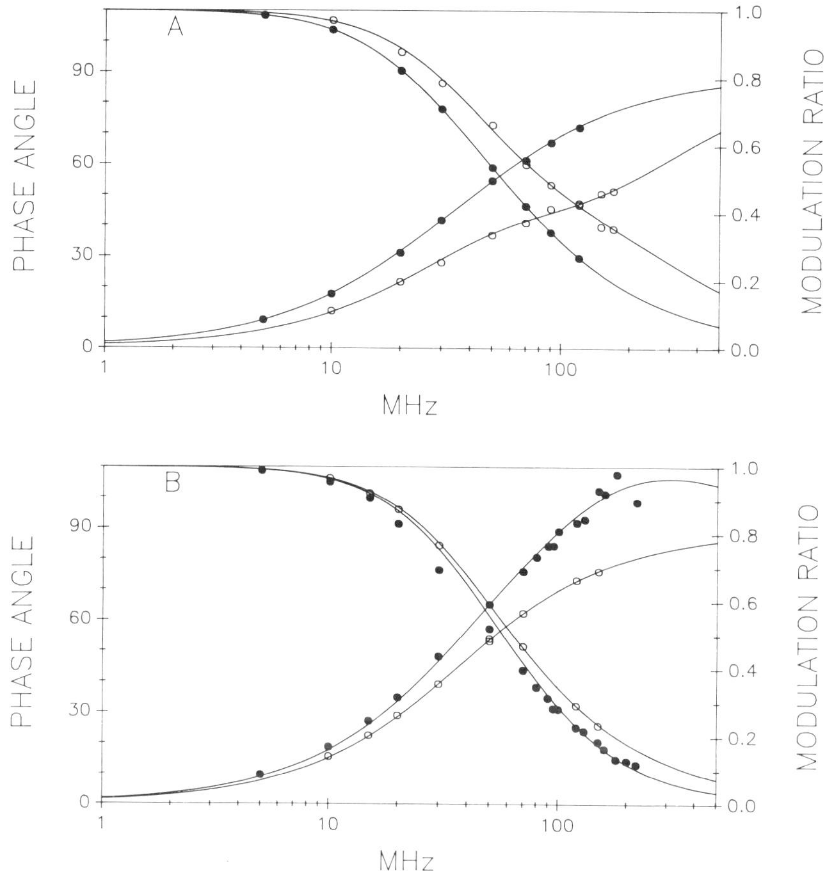

Figure 7. (A) Phase-modulation fluorescence lifetime data for the tryptophan emission of LADH (●,●) and the LADH·NADH·IB ternary complex (O,O), with excitation at 290 nm, broad band emission observed at 350 nm. (B) Phase-modulation fluorescence lifetime data for the LADH·NADH·IB ternary complex; directly excited (340 nm excitation) NADH emission (O,O), and sensitized (290 nm excitation) NADH emission (●,●).

(IB) ternary complex.[30] This enhancement (or the quenching of trp fluorescence, see below) can be utilized to experimentally obtain binding data. The fluorescence of bound NADH is blue shifted (λ_{max} = 440 nm), with respect to the NADH in water (λ_{max} = 470 nm), and the fluorescence decay of NADH in the IB ternary complex is 4.3 ns (as compared to 0.4 ns for free NADH).[26,31] Multifrequency phase/modulation data for this complex are shown in Figure 7. Only a mono-exponential decay for NADH in this complex was observed in current experimentation. Gafni and Brand[31] reported a bi-exponential decay, with a similar mean

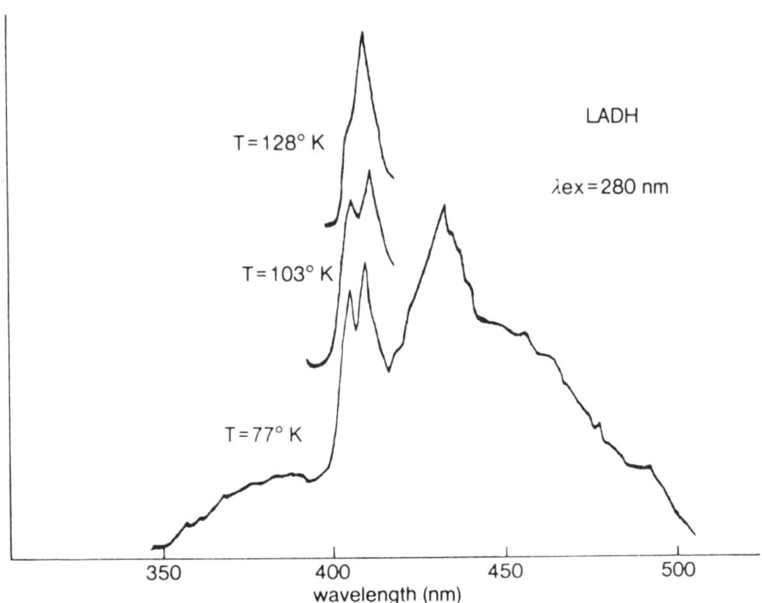

Figure 8. Phosphorescence spectra of LADH in 1:1 ethylene glycol:water glass, as a function of temperature. Reproduced from Saviotti and Galley[27] with permission from the authors and the National Academy of Sciences.

lifetime, for this complex; no explanation for this discrepancy can be found. NADH can be directly excited in the 320–340 nm range, and it can also be indirectly excited, via resonance energy transfer from the tryptophan residues (see next section).

The oxidized form of the coenzyme, NAD^+, does not fluoresce, but it does show phosphorescence at low temperature.[29] The 1,N-ethenoadenine analog of NAD^+ (ϵNAD^+) binds to LADH and has been used as both a fluorescence and phosphorescence probe for the coenzyme binding site.[29,32,33]

In addition to the natural ligands NADH and NAD^+, other extrinsic luminescence probes that bind to LADH are auramine O, 13-ethylberberine, chlorprothixene, acridine orange, and ANS.[34–37] Energy transfer studies with auramine O are discussed below.

IV. STRUCTURAL STUDIES: SPECTRAL POSITIONS, FLUORESCENCE QUENCHING, AND RESONANCE ENERGY TRANSFER STUDIES

Structural information about LADH has been provided in terms of the spectral shifts of the probes, the solvent accessibility of tryptophan residues and other bound fluorophores, and in terms of the distance between energy transfer donor and acceptor probes. These types of structural information are very low resolution, in comparison with other biophysical techniques. However, fluorescent techniques have the advantages of convenience and versatility of use with other techniques (i.e., stopped flow methods) and the advantage of being able to selectively focus on a small number of probe sites.

A. Spectral Positions of Probes

The fluorescence λ_{max} of the two trp residues clearly indicates a difference in the microenvironments of Trp-314 and -15. Trp-314 has a λ_{max} near 320 nm, which suggests a non-polar and/or non-relaxing surrounding. This λ_{max} is not among the bluest for trp residues in proteins, however. The trp of azurin has a λ_{max} of 308 nm. Although our understanding of all the factors that determine a Stokes shift in a protein is not totally clear, the suggestion is that the microenvironment of Trp-314 is not quite as non-polar as that for azurin. On the other hand, the λ_{max} = 340 nm for Trp-15 is consistent with the face of the indole ring of this residue being completely accessible to solvent.

The bathochromic absorption of Trp-314 (see Fig. 2), in comparison to Trp-15, has not been explained, but provides a means of preferentially (but not exclusively) exciting Trp-314 at longer excitation wavelengths.

These two trp residues also show different 0–0 vibronic phosphorescence transitions, with Trp-314 at 410 nm and Trp-15 at 405 nm. The reversed Stokes pattern is well documented; indole in non-polar solvents phosphoresces to the red of indole in water.[15]

The excitation and emission λ_{max} for bound NADH can be related to the microenvironment of this lumiphore, as well. Baumgarten and Hones[93] have studied the solvent dependence of the spectral properties of NADH (and its model compounds). The emission λ_{max} = 432 nm for NADH in the IB ternary complex indicates that this group is bound in a microenvironment having a polarity characterized by the Reichardt solvent polarity parameter, E_T, of 42 (e.g., polarity similar to the solvents acetone or methylene chloride).

B. Solute Quenching Studies: Fluorescence Level

Fluorescence quenching studies with iodide and acrylamide both show Trp-314 to be essentially inaccessible (i.e. $k_q^s \sim 0$).[38] We have estimated the acrylamide quenching rate constant, k_q^s (where superscript s denotes the excited singlet state), for Trp-314 to be approximately $1 \times 10^7 M^{-1}s^{-1}$. This k_q^s is an upper estimate, and, if accurate, is one of the lowest k_q^s values for acrylamide quenching. Iodide and acrylamide are charged and polar quenchers, respectively, and are known to preferentially quench surface fluorophores. Molecular oxygen is a smaller and less polar quencher. The oxygen k_q^s for Trp-314 in LADH is found to be ~1 × 10^9 M^{-1}s^{-1}.[19,20] This k_q^s is much larger than that for iodide and acrylamide, but is one of the smallest values that has been reported for oxygen quenching of a tryptophan residue in a protein.[41,43] It is generally accepted that oxygen quenches such internal trp residues by dynamic penetration through a protein matrix.[41,43,44]

Trp-15 is accessible to iodide, acrylamide, and oxygen.[16–19] With each quencher, the k_q^s is moderately large in comparison with k_q values for trp residues in other proteins. This is consistent with Trp-15 being on the surface of LADH.

The accessibility of Trp-314 and -15 does not appear to change significantly as various active site ligands are bound to the protein. For example, ternary complexes with NAD$^+$ and trifluoroethanol (TFE), or NAD$^+$ and pyrazole, show Trp-314 to be inaccessible to acrylamide and Trp-15 to be quenched with k_q^s ~ 1 × 10^{9}M^{-1}s^{-1}.[18,24,45] The same is true for binary complexes with adenosine 5′-diphosphoribose and nicotinamide mononucleotide.[18] (Acrylamide quenching of the binary complex with NAD$^+$ is complicated by the binding of acrylamide to form a ternary complex.[18] The accessibility of these two trps also is unchanged by the alkaline transition or by the selective removal of the active site zinc ion.[14] These latter forms of LADH will be discussed in Section 7, as will changes induced by urea and guanidine hydrochloride.

The general result is that the accessibility of Trp-314 and -15 remains about the same in all binary and ternary complexes. That is, the binding of ligands does not appear to lead to the exposure of Trp-314 or the burial of Trp-15. Admittedly, the relative inaccessibility of Trp-314 (i.e., its low k_q^s) makes it difficult to detect any small structural changes in terms of changes in k_q values. Temperature dependence studies of acrylamide quenching suggest that Trp-314 becomes slightly exposed as temperature is increased from 14 to 35°C, but that this is prohibited by the binding of NAD$^+$ plus TFE (see Fig. 2). Additional insights about subtle differences in the structural rigidity around Trp-314 comes from measurements of the room temperature phosphorescence decay time of this residue in various binary and ternary complexes.[28] These results will be discussed in section 5C.

Fluorescence quenching studies have also been used to investigate the solvent accessibility of bound NADH or the coenzyme analog εNAD$^+$. Gafni[33] observed that, on binding to LADH, the fluorescence of εNAD$^+$ was still able to be quenched by iodide, with a quenching rate constant, k_q^s, that is reduced by a factor of less than two from that of free εNAD$^+$. This result suggests that the ethenoadenine of the bound coenzyme analog is accessible to the solvent. However, the quenching of bound εNAD$^+$ by L-methionine, a larger quencher, was found to be reduced 10-fold from that for free fluorophore. Together with the results for iodide quenching, this indicates that the accessibility of the bound ethenoadenine group is via a channel that is limited to smaller probes. Gafni[46] also studied the quenching of bound εNAD$^+$ by the D and L stereoisomers methionine and made the interesting finding that there is chiral discrimination in the diffusional approach of the amino acid quencher to the bound probe. He found L-methionine to have almost a 3-fold larger k_q^s than the D-isomer.

C. Solute Quenching Studies: Phosphorescence Level

A number of investigations have dealt with quenching the room temperature phosphorescence of Trp-314.[27,40,44,47–53] The phosphorescence of the surface residue, Trp-15, is quenched readily by impurities, water, or protein side chains and it cannot be observed at room temperature.[27] However, because Trp-314 is so deeply buried within the protein, it phosphoresces strongly at room temperature with a decay time of 300–400 ms. The quenching of the room temperature phosphorescence of Trp-314 by oxygen was originally investigated by Saviotti and Galley.[27] Subsequently, several groups have determined the oxygen quenching rate constant, k_q^T (where superscript T denotes the triplet excited state), for LADH.[40,44,48,49,51] There is, however, a wide range of these k_q^T values, from $3.6 \times 10^7 M^{-1}s^{-1}$ by Strambini[48] to $6 \times 10^8 M^{-1}s^{-1}$ by Calhoun et al.[44] Barboy and Feitelson[40] have found intermediate values for the oxygen k_q^T for Trp-314 of LADH; a more recent report by Calhoun et al.[51] is also an intermediate value of $k_q^T \approx 1 \times 10^8 M^{-1}s^{-1}$. The disagreement between laboratories is most likely due to difficulties in measuring oxygen concentrations, and is further complicated by the photochemical depletion of oxygen that can occur when samples are illuminated.[107] The activation energy for the oxygen quenching of the triplet of Trp-314 has been determined to be 14 kcal/mole.[47] A comparison of the oxygen quenching of LADH in solution and the crystal state shows only a 30% reduction in k_q^T when the protein is in the crystal state.[49] This similarity indicates that the conformation of the protein about Trp-314 is similar for the solution and crystal states of the protein and that the crystal lattice forces do not significantly impede the phosphorescence quenching process. It seems to be the opinion of most researchers that oxygen quenches the phosphorescence (and fluorescence) of Trp-314 by a process involving penetration of this small quenching probe into the protein structure to strike the buried Trp-314. Strambini[48] has raised the possibility that phosphorescence quenching by oxygen may occur over a distance, possibly by an exchange mechanism.

Other quenchers of the phosphorescence of Trp-314 have been studied. Calhoun et al.[50] observed that several charged and neutral polar quenchers, such as NO_2^-, methyl vinyl ketone, acrylamide, and cinnamamide, are able to quench the room temperature phosphorescence of Trp-314 with k_q^T that appear to be independent of the charge and size of the quencher. These workers suggested that an unfolding of the protein, rather than a penetration mechanism, may govern the quenching of Trp-314 by these quenchers. Later, on comparison of phosphorescence quenching data for several proteins, Calhoun, Vanderkooi, Englander and coworkers[51,52] noted that there appears to be a correlation between k_q^T values and the distance that a trp residue lies beneath the protein surface. They interpreted these results in terms of a long-range electron exchange mechanism for phosphorescence quenching.

In related studies, triplet resonance energy transfer quenchers (diffusion enhanced), such as methyl orange and the protein myoglobin, have been used to estimate the depth of burial of Trp-314 from the surface of the protein.[53] These quenchers are clearly too large for a penetration mechanism to be possible, but the possibility of there being interactions of these agents with LADH may pose a problem in estimating distances.

D. Resonance Energy Transfer Studies: Trp to Coenzymes

Resonance energy transfer can, of course, also be used to estimate distances between fixed sites on a protein. With LADH, the two classes of trps, along with bound coenzyme or other bound probes, can serve as the donor–acceptor sites. Energy transfer between the trps and bound NADH has long been suggested.[30,54] In fact, the quenching of protein fluorescence by NADH (or NAD$^+$) has been widely used for binding studies.[30,55] NADH has an absorption band at 340 nm that nicely overlaps trp emission. The existence of trp→NADH energy transfer is demonstrated by both quenching of the donor (trp) emission and sensitized acceptor (NADH) emission.[39,54] Shown in Figure 4 are spectra of LADH and the LADH-NADH isobutyramide ternary complex. From the quenching of trp fluorescence, an efficiency of energy transfer, E, of 67% is obtained, in agreement with the literature.[16,17,54]

Since LADH is a dimer, there is the possibility that energy transfer can occur both between trp→NADH groups within the same subunit and between those groups on opposite subunits.[54] Listed in Table 2 are distances, from crystallographic data, between the center of the indole rings of the trp residues and the nicotinamide ring of NADH. As can be seen the Trp-314→NADH pair in the same subunit has the smallest separation distance of 17.3 Å. However, the distance between Trp-314 and the NADH in the opposite subunit is 22 Å, which is not too much longer and which is shorter than the Trp-15→NADH distance (in the same subunit; the Trp-15→NADH distance between subunits is much too large to be of importance). Theorell and Tatemoto[54] compared the ligand concentration dependence of fluorescence changes and absorbance changes and concluded that ~68% of the energy transfer quenching occurs with the binding of one NADH molecule to the dimer. This ratio of quenching by one bound ligand to that for both ligands is referred to as the parameter δ. Since this δ value is larger than 50% there is energy transfer between Trp-314 of one subunit and NADH bound to the other subunit. The efficiency of energy transfer between the three important D→A pairs (i.e., Trp-314→NADH of same subunit; Trp-314→NADH of opposite subunit; Trp-15→NADH of same subunit) is not easily assigned from steady state fluorescence studies. Time-resolved fluorescence studies from this laboratory that provide some assignments of the E values for the pairs are presented.

Table 2. Resonance Energy Transfer in LADH: Experimental Data and Crystal Structure Data[a]

Process	Fluorescence Data						X-Ray Structure		
	$k_T (s^{-1})$	d_T	$R_o(exp)$ (Å)	$\kappa^2(exp)$	$J \times 10^{16}$	$R_o^{2/3}$ (Å)	r (Å)	κ^2	Θ_T
Trp314→NADH$_A$	1.2×10^9 (E = 80%)	~0.8	22.3	0.48	56.8	23.5	17.3		
							L$_a$→	0.17	42°
							L$_b$→	0.17	43°
Trp 314→NADH$_B$					56.8	23.5	22.1		
							L$_a$→	0.58	11°
							L$_b$→	1.23	85°
Trp 154→NADH$_A$	0.075×10^9 (E=33%)	—	22.7	0.68	90.0	22.6	25.4		
							L$_a$→	0.63	32°
							L$_b$→	0.41	79°
Trp 314$_A$→Trp 314$_B$	?	~0.85	?	?	1.96	13.4	6.7		
							L$_a$→L$_a$	3.26	45°
							L$_a$→L$_b$	0.21	44°
							L$_b$→L$_a$	0.16	41°
							L$_b$→L$_b$	1.32	53°

[a]The subscripts A and B refer to the first and second subunit of LADH. The $R_o(exp)$ is calculated from k_T by the relationship $R_o(exp) = r(k_T \tau_{O,D})^{1/6}$, where $\tau_{O,D}$ is the unquenched fluorescence lifetime of the donor. $\kappa^2(exp)$ is calculated from $\kappa^2(exp) = (\frac{2}{3}(R(exp)/R_o^{2/3})^6$. The overlap integrals, J, were calculated in the standard way from normalized emission spectra and the absorption spectrum attributed to Trp-314 and Trp-15. These absorption spectra were obtained from the excitation spectra in Fig. 2 with the assumption that the molar extinction coefficients are $5 \times 10^3 M^{-1}cm^{-1}$ for tryptophan at 285 nm. For NADH as acceptor, a molar extinction coefficient of $6 \times 10^3 M^{-1}cm^{-1}$ at 330 nm was used. Quantum yields of 0.38 and 0.19 were used for Trp-314 and -15 respectively; the refractive index, n, was taken as 1.5. $R_o^{2/3}$ was calculated from J by assuming $\kappa^2 = 2/3$. The crystal structure r and κ^2 values were obtained from the structure of the LADH·NADH·DMSO complex determined by Eklund et al.[10] and obtained through the Brookhaven Protein Data Bank. Θ_T is the angle between the donor and acceptor transition moments and is related to the energy transfer depolarization factor by $d_T = (3\cos^2\Theta_T - 1)/2$. In calculations of κ^2 and Θ_T involving NADH as acceptor, the transition moment of the dihydronicotinamide ring was assigned the following direction, based on the work of Evleth et al.[96] Likewise, the assumed orientations of the ^1L_a and ^1L_b of indole are shown below.

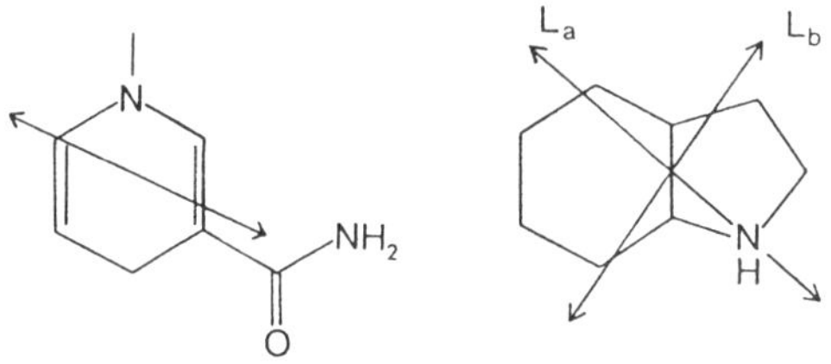

Multi-frequency phase/modulation data for trp fluorescence of LADH and its NADH-isobutyramide ternary complex are shown in Figure 7. Also shown in Figure 7 are data for the directly excited (340 nm) and indirectly excited (sensitized, 290 nm) emission of bound NADH. These data show the dynamic quenching of trp emission caused by coenzyme binding. Bi-exponential decay fits give a quenching of the long-lived (Trp-15) component from 7.2 ns in the free enzyme to 4.5 ns in the ternary complex. For the short lived (Trp-314) component, the decay time drops from 3.6 ns to 0.68 ns on forming the ternary complex. Thus both trp residues appear to be dynamically quenched, with much greater quenching of Trp-314, as expected from its closer distance. (Note that with the conditions employed, the protein was ~100% saturated with NADH. It turns out that energy transfer between the closest Trp-314→NADH pairs, those being on the same subunit, will be more significant than that between the subunits. Only when the degree of saturation is much less than 100% will transfer between the subunits be kinetically significant.)

Regarding the directly excited NADH emission, we find a mono-exponential decay time for the bound coenzyme. For the sensitized emission, the multi-frequency phase/modulation data have an unusual appearance, as shown in Figure 7B. A phase angle $>90°$ is measured. This result is diagnostic of the existence of an excited state reaction (i.e., resonance energy transfer).[56]

We have performed a global analysis[57] of these four sets of data in Figure 7 to a kinetic model that includes energy transfer between two D→A pairs, Trp-314 to NADH and Trp-15 to NADH. (Again note that transfer from Trp-314 to NADH on the opposite subunit will be a minor kinetic route when the protein is saturated with NADH.) The fit, shown in Figure 7, yields energy transfer rate constants, k_T, of $1.21 \times 10^9 s^{-1}$ and $0.075 \times 10^9 s^{-1}$ for Trp-314 and Trp-15 as donors. These correspond to E values of 80% and 33%, respectively.

As mentioned in the Introduction, since the D→A distances are known for this system from the extensive crystal structure data, these energy transfer measurements enable testing of other aspects of the process, such as the orientation factor, κ^2. This orientation factor can theoretically range from 0 to 4.0, but in most applications of resonance energy transfer, researchers assume the dynamic-limit isotropic value of $\kappa^2 = 2/3$; that is, it is assumed that the D and A oscillators can rapidly rotate to achieve all mutual orientations within the lifetime of D. In this study with LADH the experimental E values (or rate constant for energy transfer, k_T, values) was used to solve for the actual critical distance, R_o. This experimental R_o, together with the $R_o^{2/3}$ (calculated assuming $\kappa^2 = 2/3$) that is determined from the spectral overlap integral, J, can be used to calculate the experimental orientation factor, κ^2, as $\kappa^2(exp) = (2/3)(R_o(exp)/R_o^{2/3})$. The experimental κ^2 can then be compared to values determined from inspection of crystal data to test the accuracy of the method in estimating distances and to possibly reveal instances in which differences in orientation factors allow the assignment of which electronic oscillators (i.e., ^1L_a or ^1L_b of indole) are involved. In Table 2 are listed the experimental k_T, $R_o(exp)$ and $\kappa^2(exp)$ values for Trp-314→NADH and Trp-15→NADH energy transfer. (See the legend of this Table for the explanation of how the spectral overlap, J, was determined.) Fluorescence anisotropy decay measurements for trp and bound NADH emission (see following section) show that Trp-15, Trp-314 and NADH are each relatively immobilized in the protein structure. Thus the κ^2 values should be determined by the mutual orientation of electronic transitions.

If the solution structure is the same as the crystal structure, these orientations and the D→A separation distance, r, are determinable from inspection of the latter. However, there is still ambiguity since indole has two nearly degenerate and orthogonal electronic oscillators, ^1L_a and ^1L_b, and it is not always clear which serves as the donor. For the solvent exposed Trp-15, it is most likely that the ^1L_a oscillator (which lies between the indole's C4–C5 midpoint and the nitrogen of indole) is the lowest excited singlet state; the ^1L_a transition is usually the lowest energy singlet state for indole derivatives in polar solvents.[58,59] For Trp-314 it is not clear whether ^1L_a or ^1L_b will serve as the donor, since the ^1L_b (which lies between C7 and the C3–C9 midpoint of the indole ring) can be the lowest energy state in apolar media. Listed in Table 2 are values of κ^2 obtained from the crystal structure for the various possible transitions. The experimental (fluorescence) κ^2 are in good agreement with the values expected from the crystal structure, whether the ^1L_a or ^1L_b transition serves as the donor for either Trp-15 or Trp-314. The $\kappa^2 = 0.17$ found for Trp-314→NADH (for both ^1L_a and ^1L_b as donor, coincidentally) from the X-ray data is actually not significantly different from the experimental $\kappa^2(\mathrm{exp}) = 0.48$, when one considers that the $\frac{1}{6}$ root of these values is what is important. Another way to look at this study is that the $R_o(\mathrm{exp})$ and $R_o^{2/3}$ are all fairly similar in this system, thus permitting the usual assumption that $\kappa^2 = \frac{2}{3}$ with little error, even though neither the donor or acceptor undergoes rapid, isotropic rotation in this system. We do not wish to imply that κ^2 can always be taken as $\frac{2}{3}$ in such "rigid" protein systems (since if $\kappa^2 << 0.1$, significant errors can result). This exercise is intended only as an examination and comparison of energy transfer theory with crystallographic data.

The binding of NAD^+, plus TFE or pyrazole, also quenches trp fluorescence, particularly that of Trp-314.[16,17,22,24,54] The maximal extent of quenching is about 50% for bound NAD^+ in the TFE ternary complex (as compared to 70% quenching for bound NADH). The mechanism of this quenching has been suggested to be either 1) an induced conformational change that brings a quenching group (i.e., methionine side chain or peptide bond) closer to Trp-314,[17,54] or 2) resonance energy transfer to bound NAD^+.[17,22] Although the overlap integrals are small, Abdallah et al.[17] and Ross et al.[22] have calculated energy transfer efficiencies of 35% and 38% for transfer from Trp-314 to the adenine and nicotinamide moieties of NAD^+, respectively. The degree of quenching is greater in the TFE ternary complex (~50%) than in the pyrazole ternary complex (~35–40%). This difference is difficult to rationalize in terms of the energy transfer mechanism, since the overlap integral is actually larger for the pyrazole complex. This fact is illustrated by the difference spectra in Figure 9, which show that the NAD^+·pyrazole ternary complex has a much stronger absorbance at 300 nm and above, as compared to the NAD^+·TFE ternary complex.[60,61] Slight differences, between the two ternary complexes, in the mutual orientation of Trp-314 and the nicotinamide (or adenine) ring must be postulated. (Also, the phosphorescence of Trp-314 is almost totally quenched in the NAD^+·TFE ternary complex, but not in the NAD^+·pyrazole ternary complex. The reason for this difference is not certain.) Since nicotinamide and N-methylnicotinamide salts act as diffusional quenchers of the fluorescence[62] and

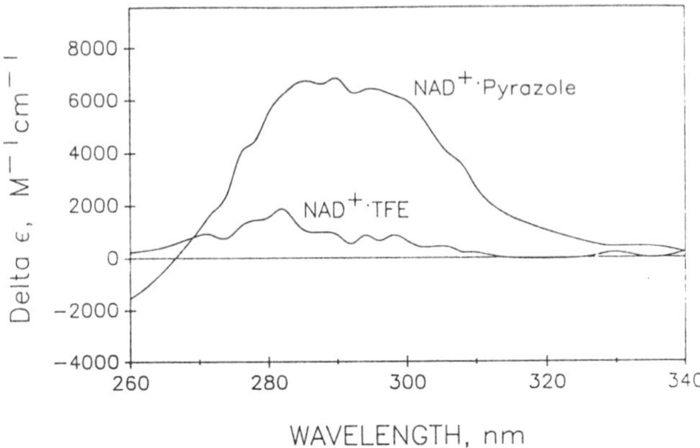

Figure 9. UV differences spectra for the LADH·NAD⁺·TFE and LADH·NAD⁺·pyrazole ternary complexes. Spectra are that of the mixture minus the components. Conditions, pH 7.2, 0.03 M sodium phosphate, 20 °C.

phosphorescence[51] of tryptophan, probably by an electron transfer mechanism, it should be considered that bound NAD⁺ may quench Trp-314 by electron transfer through the protein fabric.[28,29]

The room temperature and low temperature phosphorescence of Trp-314 of LADH is also quenched by the binding of coenzyme; for NADH·isobutyramide and NAD⁺·pyrazole ternary complexes the quenching appears to occur at the singlet level and involves resonance energy transfer.[28] The phosphorescence of the NAD⁺·TFE complex is almost completely quenched.[28] There must be an additional quenching mechanism in this complex that operates at the triplet level in the latter ternary complex. Since NAD⁺ acts as a diffusional quencher of room temperature trp phosphorescence, Strambini and Gonnelli[28] have suggested that the bound NAD⁺·TFE center may quench the triplet state by an intermolecular electron transfer process. The observation that the NAD⁺·pyrazole complex does not show this additional phosphorescence quenching mechanism gives further support to the notion that the nicotinamide group in the latter complex is "NADH-like" due to the formation of a partial bond between C4 of NAD⁺ and one of the nitrogens of pyrazole.[28,60]

Strambini and Gonnelli have argued that the degree of phosphorescence quenching of LADH binary and ternary complexes, whether it be due to energy transfer from the singlet level or to a new quenching mechanism at the triplet level, is thus a sensitive means of probing for subtle differences in the structure of various complexes. For example, most NADH containing complexes (quenched by energy transfer from the Trp-314 singlet level to the dihydronicotinamide) show a residual phosphorescence of 15%, as compared to LADH. The ternary complex with NADH and DMSO, however, shows 20% residual phosphorescence, leading to the conclusion that the D→A distance or mutual orientation is less favorable for energy transfer in this complex than in others.

E. Resonance Energy Transfer Studies: Trp to Other Probes

Resonance energy transfer also can occur between the trps of LADH and other bound extrinsic probes. The cationic dye auramine O, AO, has a very weak fluorescence in aqueous solution, but, on binding to LADH, its fluorescence intensity increases 1000-fold.[35,63] AO appears to bind adjacent to the alcohol binding site. Not only is the fluorescence of AO enhanced upon binding, but it quenches the trp fluorescence of the protein. Using wavelength and time-resolved measurements, Brand et al.[64] demonstrated that AO primarily quenches Trp-314. The overlap between trp emission and AO absorbance, and the demonstration of sensitized fluorescence of AO, indicates that resonance energy transfer occurs. As with the above case with bound NADH,[54] Brand et al. found that quenching was not linear with respect to coverage of binding sites, or, in other words, the Trp-314 residues from both subunits are able to transfer to the first AO that binds to one of the subunits.

Triplet-singlet energy transfer from Trp-314 to bound AO also occurs, as shown by Weers and Maki[63] using optically detected magnetic resonance (ODMR) spectroscopy and by measuring delayed luminescence of bound AO in a glass at 77 °K. From analysis of the kinetics of the decay of the AO delayed luminescence these workers obtained two rate constants for triplet-singlet energy transfer from Trp-314 to AO. Several explanations were considered, including the possibility that two distinct conformations of the LADH·AO complex exist, with different orientation factors or separation distances between the donor and acceptor, under the experimental conditions. From ODMR studies, the rate constants for energy transfer from individual Trp-314 triplet state spin sublevels were also determined.

F. Resonance Energy Transfer Studies: Trp to Trp

The existence of homotransfer between trp residues in proteins has been often discussed.[65–67] For LADH, the closeness of the two Trp-314 residues ($r = 6.7$ Å separation between ring centers in the NADH·DMSO ternary complex), the blueness of the emission, the redness of the absorption, and the high quantum yield of Trp-314 makes this a likely system for homotransfer. The Forster critical distance, R_o, depends on the spectral overlap integral, J, the donor quantum yield, Φ_D, the orientation factor, κ^2, and the index of refraction, n, of the medium between the donor and acceptor. According to calculations by Eisinger et al.,[67] R_o for trp→trp transfer in proteins may typically be in the range of 5.3–11.8 Å. However, Trp-314 has an exceptionally high Φ_D and favorable J; the $R_o^{2/3}$ for homotransfer between the Trp-314's in LADH (assuming $\kappa^2 = 2/3$) is estimated to be a relatively high value

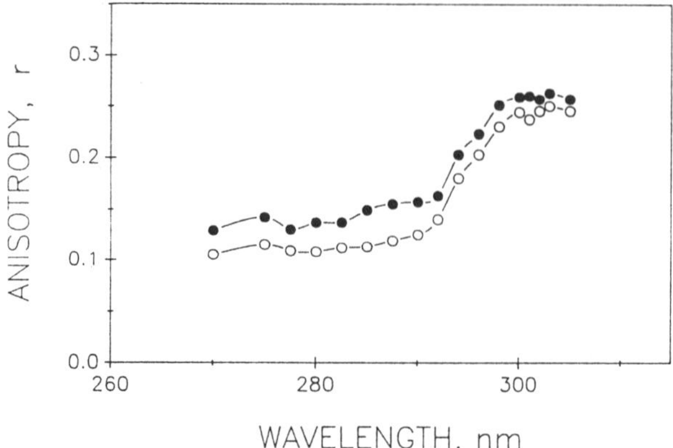

Figure 10. Fluorescence excitation anisotropy spectra of NADH in the absence (O) and presence (●) of 0.55 M KI. Conditions, pH 7.2, 0.03 M sodium phosphate, 0.55 potassium chloride (or iodide), 20 °C, emission observed at 323 nm.

of 13.4 Å. Depending on whether the actual κ^2 is larger or smaller than $\frac{2}{3}$, the actual R_o will likewise be larger or smaller than 13.4 Å, respectively. Since the actual r between the two Trp-314 of LADH is much smaller than the expected R_o, transfer between these two residues is therefore anticipated (see the comment below on other considerations for small r).

Other than the proximity of the two Trp-314 residues, however, there is no experimental evidence for their homotransfer. Admittedly, such evidence is difficult to obtain. One way to demonstrate homotransfer is to observe a depolarization of the emission due to the loss of anisotropy that would accompany energy transfer, provided that the oscillators are not parallel. In Figure 10 is shown the anisotropy excitation spectrum of LADH in the absence and presence of 0.55 M KI (the latter being added to selectively quench most of the fluorescence of Trp-15 and to therefore simplify the spectrum). The anisotropy remaining in the presence of quencher is relatively large at all wavelengths. The ratio r_{300}/r_{270} (i.e., the anisotropy at 300 nm, where the "red edge effect" limits homotransfer,[68] divided by the r at 270 nm) is ~2.0 for Trp-314. This r_{300}/r_{270} ratio of 2.0 is only slightly larger than that for immobilized N-acetyltryptophanamide ($r_{300}/r_{270} = 1.66$)[65] and thus indicates that there is little depolarization of Trp-314 fluorescence due to homotransfer (i.e., the depolarization factor for energy transfer, d_T, is estimated to be about 0.85). From these results it can be concluded that either 1) homotransfer does not occur to a significant extent or 2) the electronic oscillators for the two Trp-314 are nearly parallel, so that little depolarization occurs with energy transfer.

From the crystal structure of LADH the expected κ^2 and d_T factors for Trp-314 homotransfer can be calculated. Considering the directions of the ^1L_a and ^1L_b transition moments of indole, the four possible κ^2 and d_T (listed as the azimuthal angle, Θ_T, where $d_T = (3\cos^2\Theta_T\ 1)/2$ values listed in Table 2 are calculated. The values of κ^2 range from 0.16 to 3.26. None of these are sufficiently close to zero to preclude their participation in energy transfer. The $^1L_a\rightarrow{}^1L_a$ transfer process has a very favorable κ^2, and, since the ^1L_a absorption often extends to the red of the sharp 0–0 ^1L_b band, it appears likely that the $^1L_a\rightarrow{}^1L_a$ energy transfer process would be the most efficient. The Θ_T value of 45° for these two oscillators gives a transfer depolarization factor, d_T, of 0.25, which is much larger than the experimentally estimated d_T value of 0.85. However, since the two Trp-314 are identical, energy transfer between the two should be reversible, which would have the effect of essentially reducing the effective Θ_T in half, corresponding to a d_T of 0.78. This latter value is near that estimated from the data in Figure 10. Given the empirical nature of the argument relating to Figure 10 and the uncertainly about ^1L_a and ^1L_b transition directions, this analysis is consistent with there being facile energy transfer between the two Trp-314s.

It should be noted that due to the small distance between the two Trp-314 (3.3 Å between closest ring carbons), the exact calculation of the rate constant for homotransfer should consider dipole–quadropole interactions, in addition to the dipole–dipole interaction, and possibly should also consider the exchange interaction.[67] In fact, we speculate that the redness of the absorption spectrum of the Trp-314 residues might be caused by the mutual interaction between their molecular orbitals.

V. STRUCTURAL DYNAMICS: FLUORESCENCE QUENCHING, PHOSPHORESCENCE LIFETIMES AND ANISOTROPY DECAYS

Information about conformational fluctuations, at the probe sites, can be obtained by certain luminescence quenching methods, anisotropy decay measurements, and phosphorescence lifetime measurements. Fluctuations can further be characterized, using these methods, by studying the effect of temperature, comparing different complexes, and comparing the solution and crystal states of the protein. As will be seen, the various luminescence methods provide different views of the dynamic character of LADH.

A. Solute Quenching of Fluorescence and Phosphorescence

Fluorescence quenching by oxygen shows that the k_q^s is ~1 × $10^9\mathrm{M}^{-1}\mathrm{s}^{-1}$ for Trp-314.[18] This is among the smallest for trp residues in proteins.[41–43] Quenching by oxygen apparently involves collision between the quencher and the excited indole and quenching of Trp-314 must involve penetration of oxygen through the protein to reach Trp-314. The accepted view is that this penetration is facilitated by small amplitude fluctuations in the protein structure to form a transient diffusion path for the small quenching probe.[41–44] The k_q^s for oxygen quenching of Trp-314 is only 10-fold reduced from that for the collisional quenching of indole in water. Thus, the penetration by oxygen is rather facile and the microviscosity around Trp-314 is suggested by this method as being 10-fold larger than that for water. There has been very little work done to compare the oxygen fluorescence quenching of Trp-314 in different complexes of LADH;[20] neither has there been temperature dependence studies. This is due to the difficulty in resolving the fluorescence of Trp-314 and Trp-15 (which requires time-resolved measurements) and the need to use a cumbersome high pressure cell to achieve the >1000 psi of O_2 pressure required to observe a significant degree of quenching.

Fluorescence quenching by acrylamide turns out not to provide much information about the dynamics of LADH, since Trp-15 is quite accessible and Trp-314 is essentially inaccessible to this quencher. From temperature dependence studies, a slight amount of quenching of Trp-314 appears as temperature is increased from 25–35°C.[24] For ternary complexes with NAD^+ plus TFE or pyrazole, this slight quenching at higher temperature is not seen. Of course, these data must be interpreted with caution, since they may be reflecting a difference in the thermal unfolding of the free protein and the complexes.

Oxygen and other molecules are also able to quench the room temperature phosphorescence of LADH. As mentioned in section 4C, only Trp-314 contributes to the room temperature phosphorescence of LADH, simplifying such quenching studies.[27] Several groups have measured the k_q^T for oxygen quenching of Trp-314,[27,40,44,48,49] with values ranging from 8×10^6 to 6×10^8 $M^{-1}s^{-1}$. No comparative studies have been made of the oxygen k_q^T for various ternary complexes of LADH. Since other phosphorescence studies (see below) suggest that the microenvironment of Trp-314 becomes less flexible on coenzyme binding, such oxygen k_q^T values would be expected to decrease with ligation of the protein.

B. Anisotropy Decays Fluorescence and Phosphorescence

Anisotropy decay measurements have proved to be very useful for characterizing internal motions within proteins.[69,70] Frequently it is found that trp residues will experience very rapid (subnanosecond) wobbling motion about their C_α—C_β and/or C_β—C_γ bond, in addition to experiencing the global rotation of the globular protein. Since the fluorescence decay of trp occurs in the 1–10 ns range, it is well suited for revealing such motions on this time scale. For LADH, the only reported anisotropy decay measurement, of trp fluorescence, shows a single rotational correlation time, ϕ, of 36 ns at 27.5°C, to adequately describe the rotational motion of both Trp-15 and Trp-314.[22] No rapid, wobbling motion was revealed for either residue. Quenching resolved emission anisotropy studies support this result.[42,71] Similarly, a single rotational correlation time has been reported for the fluorescence of bound NADH and the bound probes, ANS and TNS.[104] Beechem et al.[104] have simultaneously analyzed anisotropy decay data for these bound probes, along with data for trp fluorescence, according to a model in which the protein is assumed to be an anisotropic rotator (prolate ellipsoid) and the various probes are assumed to bind at different angles with respect to the equatorial (short) axis and polar (long) axis of the protein. The anisotropy decay data for each probe thus senses, to different degrees, the rotational diffusion about these two major axes. The results of this analysis describe LADH as being an ellipsoid of axial ratio 3.1. This value is actually very close to the maximum axial ratio of 3.0 that is obtained from the crystal structure. This exemplifies the use of anisotropy decay measurements to provide information about the overall hydrodynamic properties of a protein.

Phosphorescence anisotropy decay measurements, at room temperature, have also been reported for the Trp-314 of LADH.[72] Of course the long phosphorescence decay time of ~400 ms allows for complete rotational depolarization of Trp-314 luminescence for the protein in solution. However, the phosphorescence anisotropy decay of crystalline LADH, for which global rotation is prohibited, has been studied by Strambini and Gabellieri.[72] These researchers found essentially no decay of the phosphorescence anisotropy of Trp-314, indicating a lack of independent rotational motion of this residue's indole ring on the seconds time scale. An estimate was made that any rapid wobbling motion of the indole side chain must be of limited amplitude (i.e., within a cone of semi-angle of less than 15°). It was also estimated that the rigidity of the microenvironment around Trp-314 can be described in terms of an effective viscosity of greater than 2.5×10^8 P.

C. Phosphorescence Lifetimes

A third luminescence method that seems to reveal information on the flexibility of LADH and its complexes is measurement of the Trp-314 room temperature decay times. A correlation has been noted between phosphorescence lifetimes and the degree of burial and rigidity of trp residues.[73,74] This relationship has been modeled by studies with indole in solutions of different viscosity, where a correlation between the triplet lifetime and viscosity (or rotational correlation time) was observed.[73] The ~400 ms lifetime of Trp-314 suggests a local microviscosity of around 10^5 P. In a study of a dozen binary and ternary coenzyme complexes with LADH (ones where triplet level quenching appears not to occur), the phosphorescence lifetime was found to vary from 200 to 1900 ms, presumably reflecting differences in the rigidity of the Trp-314 microenvironment in the various complexes.

The different luminescence methods thus each portray the protein environment around Trp-314 to be relatively inflexible. There is qualitative but not quantitative agreement between oxygen quenching, phosphorescence lifetime, and anisotropy decay measurements. This has been discussed by Strambini and Gabellieri.[72] Since the different methods probably involve different types and amplitudes of molecular displacements, agreement cannot be expected. Quenching by oxygen requires the formation of rather small channels for this 2–3 Å probe, whereas depolarization of the phosphorescence anisotropy would require the displacement of the surrounding protein chain and side groups to enable the out-of-plane rotation of the indole ring. Since Trp-314 is tightly embedded between two β-sheets, such ring flipping motions may be strongly prohibited.

VI. THERMODYNAMICS AND KINETICS OF LIGAND INTERACTIONS

A substantial review of the thermodynamics and kinetics of the interaction of LADH with coenzyme, substrates, and inhibitors has been given by Pettersson.[13] In this section the role that luminescence measurements has played in these studies will be discussed.

The ~13-fold enhancement of the fluorescence of NADH upon binding was used by Theorell and McKinley-McKee[30] to develop a convenient method for determining binding isotherms for NADH to LADH; this general procedure has been applied to many other dehydrogenases.[55] In the presence of isobutyramide the binding of NADH is very strong ($K_{diss} \approx 5 \times 10^9 M^{-1}$) and thus the NADH enhancement procedure can also be used as an active site titrant to quantitate the concentration of enzyme.[30]

The quenching of trp fluorescence by NAD^+ binding is a second fluorimetric technique for determining association constants.[75,76] In view of what is known about the quenching of trp fluorescence by NADH (51, see section 4D), it is not clear a priori whether or not the quenching on NAD^+ binding is linear in the coverage of sites. That is, does the binding of NAD^+ to one subunit quench Trp-314 fluorescence of the other subunit, as well as that of the first subunit? To answer this question a careful analysis of trp quenching–binding profiles, in terms of the signal change for each step, can be performed. For example, Anderson and Dahlquist[76] analyzed data for NAD^+ binding to LADH by fitting for two macroscopic association constants (i.e., for the binding of the first and second ligands) and two step-wise fluorescence signal changes. Their analysis yielded an equal degree of quenching for binding to the first and second sites and also yielded values for the two association constants.

The question of the interaction between the two subunits of LADH has been addressed and answered differently by various workers.[75,79–81] The above mentioned data and analysis of Anderson and Dahlquist[76] gave Hill coefficients, for binding of NAD^+ plus TFE or pyrazole to the two subunits, of 1.16 and 0.89, values which are essentially equal to 1.0. This indicates that there is no significant site–site interaction in LADH, at least with respect to coenzyme binding. While a site–site interaction appears not to be significant, there is a strong positive heterotrophic cooperativity between NAD^+ and TFE, between NAD^+ and pyrazole, and between NADH and IB in binding to LADH.[13,30,76,77] Shown in Figures 11A and 11B are

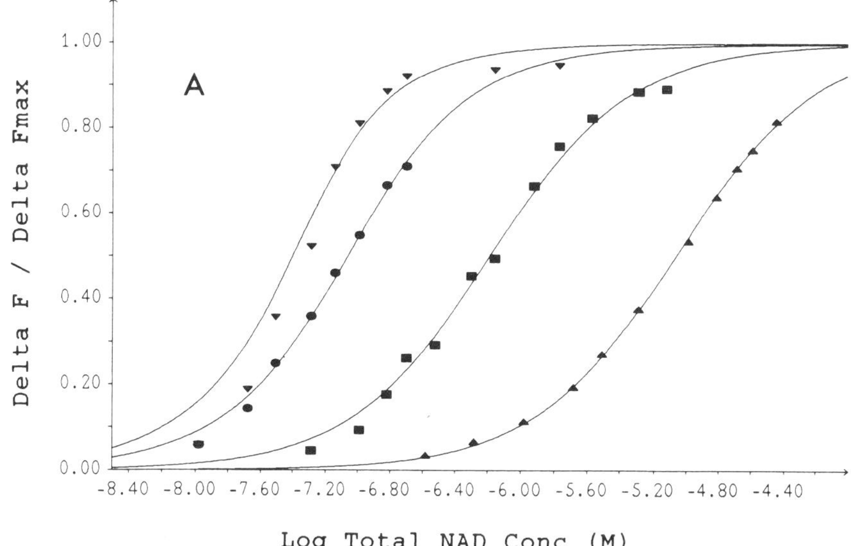

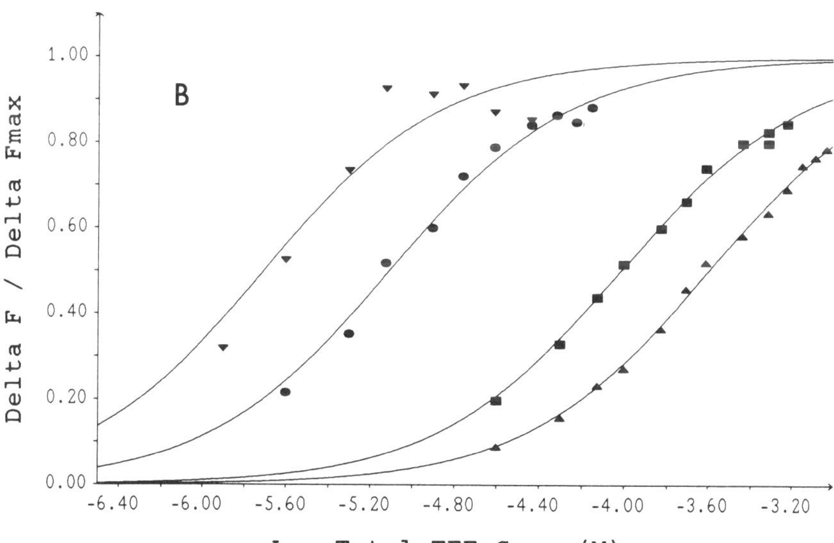

Figure 11. (A) Binding isotherms for the interaction of NAD$^+$ with LADH as a function of TFE concentration at pH 9.5, 20 °C. TFE concentrations are 0 (▲), 2.5×10^{-4}M (■), $2.3 \times) 10^{-3}$M (●), and 5×10^{-2}M (▼). (B) Binding isotherms for the interaction of TFE with LADH as a function of the concentration of NAD$^+$ at pH 7.5, 20 °C. NAD$^+$ concentrations are 5×10^{-7} (▲), 1.25×10^{-6}M (■), $2.5 \times) 10^{-5}$M (●), and 5×10^{-4}M (▼).

binding profiles for NAD⁺ as a function of TFE concentration and binding profiles for TFE as a function of NAD⁺ concentration.[77] These were obtained in our laboratory by measuring the quenching of trp fluorescence on binding. For the TFE binding profiles at high [NAD⁺], advantage is taken of the small difference in the fluorescence intensity of the LADH·NAD⁺ binary complex and the LADH·NAD⁺·TFE ternary complex.

Fluorescence has also been extensively used in studies of the kinetics of LADH-coenzyme interactions and catalytic reactions.[60,82,83] The fluorescence change upon reduction of NAD⁺ to NADH is easily monitored. Also, the quenching of trp fluorescence on NAD⁺ binding has been used with stopped flow and pressure relaxation techniques to reveal the kinetics of the diffusional binding step and a subsequent isomerization step, attributable to an induced conformational change in the protein.[60,82,103]

In a very unique approach to the question of site–site interactions, Strambini and Gonnelli[84] have measured the room temperature phosphorescence decay time of Trp-314 as a function of the degree of coenzyme (i.e., NADH plus isobutyramide or NAD⁺ plus pyrazole) binding to LADH. They found that, upon binding the first coenzyme molecule to the dimeric protein, the phosphorescence decay time increases from 0.4 s to 1–1.3 s. Such an increase suggests that the microenvironment of Trp-314 becomes less flexible as the first ligand binds. Upon binding of the second ligand, the decay time drops to 0.63 s (for NADH plus isobutyramide), which is still larger than the unliganded protein but smaller than the half-saturated complex. Since Trp-314 is located at the intersubunit interface, its phosphorescence decay time appears to be sensitive to any subtle conformational change induced by ligand binding. The variation in the decay time with degree of saturation, however, was consistent with no thermodynamic linkage between the two binding sites.

VII. THERMODYNAMICS AND KINETICS OF CONFORMATIONAL TRANSITIONS

The use of fluorescence to study pH and denaturant induced conformational changes in LADH dates back to the work of Brand and coworkers.[14] Since that time most attention has been paid to the acidic and alkaline transitions of the protein.

At pH values below 5.0 the protein slowly loses its Zn ions and undergoes a structural transition that is observed as dynamic quenching of both trp residues.[23] Using a high repetition rate laser source with a single-photon-counting instrument, Brand and coworkers[85a] have measured the fluorescence decay of LADH as a function of time (decay profiles measured over periods of two seconds for the first 11 minutes of the reaction) to monitor changes in the fluorescence lifetimes of Trp-314 and -15 as the protein undergoes the acid induced transition. Global analysis of the decay profiles was performed in terms of kinetic model in which

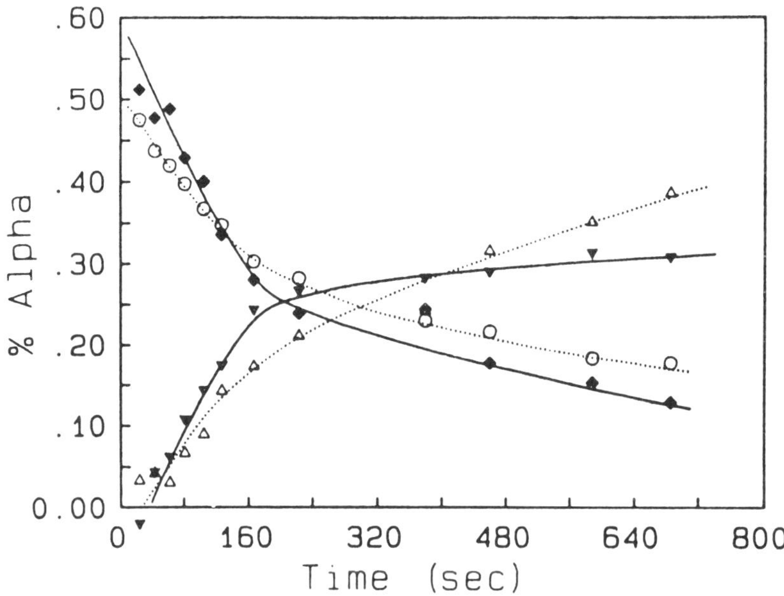

Figure 12. Kinetic decay data for the acid induced transition of LADH at 7.5 °C. Shown are the time dependent changes in the pre-exponentials associated with the native lifetimes of 3.3 ns (♦) and 6.6 ns (o) and the unfolded state lifetimes of 1.9 ns (△) and 5.1 ns (▼). Reproduced from Han et al.[100] with permission of the authors and Plenum Press.

the bi-exponential decay of the native protein (τ_1 = 3.3 ns for Trp-314 and τ_2 = 6.6 ns for Trp-15) is converted to a bi-exponential decay of the acid unfolded form of the protein (τ_1 = 1.9 ns, τ_2 = 5.1 ns). Shown in Figure 12 is the time course of the reaction in terms of the decrease in the amplitudes for the native components and the increase in the amplitudes for the unfolded components.[99,100] Recently, Beechem[105] has improved upon this methodology by use of a multi-anode microchannel plate detector, which enables simultaneous measurements to be made at multiple emission wavelengths.

The alkaline quenching of LADH occurs with an apparent pK_a of 9.5–10.0 and results in a red shift in the protein's emission.[16,85–87] A selective alkaline quenching of Trp-314 is thus indicated.[16] Time-resolved fluorescence studies support this view by showing a selective, dynamic quenching of the 3.5 ns (at pH 7) component to a value of 2.0 ns as the pH is increased to 10.5.[24] Laws and Shore[88] proposed that this selective quenching of Trp-314 is due to the acid dissociation of a nearby tyrosine residue, Tyr-286, with quenching of Trp-314 then occurring by energy transfer to the tyrosinate. The mechanism of the alkaline quenching effect and the identity of the quenching group has been an important topic based on 1) the fact

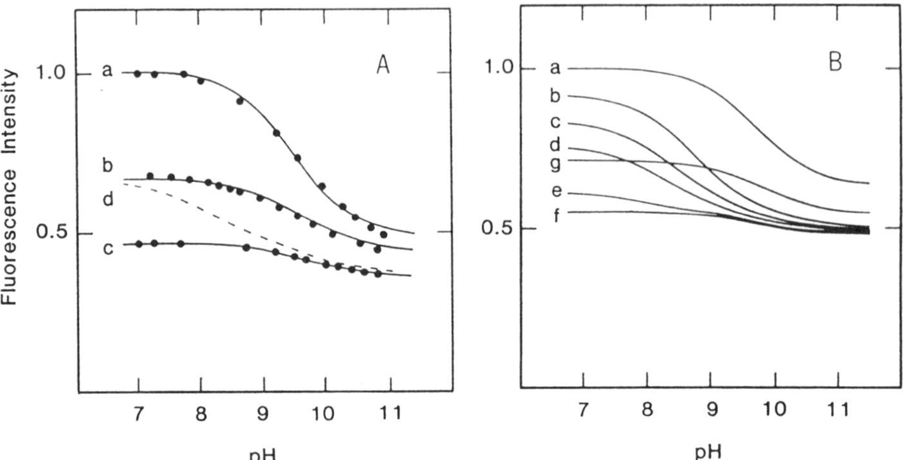

Figure 13. (A) pH dependence of the fluoregcence of LADH (a), the LADH·NAD$^+$·pyrazole ternary complex (b), and the LADH·NAD$^+$·TFE ternary complex (c). Conditions: 25 °C, 0.1 M NaCl, λ_{ex} = 295 nm, λ_{em} = 323 nm. The solid lines through (a) (b) and (c) are with pK$_a$ values of 9.5, 9.5, and 9.6, respectively. (B) Simulated pH titration curves for LADH in the absence and presence of NAD$^+$, using a model involving dynamic, competitive quenching by the alkaline transition and by NAD$^+$ binding, and with a pH dependent association constant for NAD$^+$. The curves a→e are for increasing concentrations of NAD$^+$. Curves f and g are simulated titration curves for the TFE and pyrazole ternary complexes, respectively. See reference 84 for details. Reproduced with permission of the American Chemical Society.

that the binding of NAD$^+$ also is linked to the proton dissociation from some group with pK$_a$ ~ 9.5,[12,85] and 2) the observation of a downward shift in the apparent alkaline quenching pK$_a$ when NAD$^+$ and TFE are bound.[85,86] These observations imply that the alkaline fluorescence quenching effect provides a means of characterizing the pK$_a$ of a group that is linked to coenzyme binding. For example, it has been suggested that the ionizing group is a water molecule ligated to the active site zinc and that the alkaline fluorescence quenching is due to a conformational change that occurs following the ionization of this group.[85] Our group has argued that it may just be a coincidence that the pK$_a$ of the group (i.e., Tyr-286) responsible for quenching Trp-314 is similar to the pK$_a$ of the group linked to coenzyme binding.[87] Since both the alkaline effect and NAD$^+$ binding separately cause a dynamic quenching of Trp-314 fluorescence, the two quenching processes will compete with one another, rather than being additive quenching effects. In Figure 13 is shown fluorescence pH titrations of free LADH and its NAD$^+$·TFE ternary complex. The solid lines are fits with a model in which there is competitive, dynamic quenching of Trp-314 by the two independent mechanisms, with the pK$_a$ for the alkaline transition remaining at 9.5, regardless of whether or not NAD^+ is bound. The point is that this model does not identify the alkaline quenching pK_a as being the ionizing group which controls coenzyme binding.

The effects of urea and quanidine·HCl (Gu·HCl) on the structure of LADH have recently been studied via room temperature phosphorescence measurements.[89,90] Gu·HCl, at concentrations just below those causing denaturation, was found to induce an increase in the flexibility of the Trp-314 microenvironment, as indicated by a decrease in phosphorescence decay time. A similar effect was not seen for urea. The kinetics of the urea and temperature induced denaturation of LADH has also been studied using phosphorescence. The denaturation process is found to follow a two-state model and to have a very steep urea concentration dependence.[89]

Removal of both the structural and active site zinc ions of LADH results in a marked destabilization and inactivation of the protein.[91] A selective removal of the active site zinc has been achieved.[92] The resulting apoprotein is inactive but still retains the structure of the parent protein.[92] Its coenzyme binding properties are slightly altered from those of the parent protein. For example, the zinc deficient apoprotein has an increased affinity for enzymes and fluorescence probes, such as auramine O, yet cooperativity between the binding of NADH and IB is greatly diminished.[36,92] Also, the intrinsic trp fluorescence properties of the active-site-zinc-depleted apoprotein appear to have changed very little.[24,92] Upon removal of both the structural and active site zincs, however, the fluorescence properties of the protein are changed more dramatically.[106]

VIII. FINAL COMMENTS

The above described studies with LADH stand as examples of how various luminescence techniques can be used to obtain useful information about a protein, its structure and dynamics, interactions, and transitions. LADH may be an ideal multi-tryptophan protein for these studies and similar success with other proteins may not be achievable. Nevertheless, the application to LADH has been fruitful and in the future we will probably continue to see LADH used as a "test system" for the development of new luminescence approach to characterize and resolve components in proteins. Studies with forms of the protein lacking either Trp-15 or Trp-314 would be of obvious interest to absolutely confirm many of the interpretations presented in this chapter. There has been a report of a procedure for the selective chemical modification of Trp-15.[101] However, we have been unable to achieve this selective modification in our laboratory. No reports of site directed mutagenesis of this protein have yet been made. Such studies are eagerly anticipated, since the selective removal of the individual trp residues would allow the assignments of luminescence properties to be confirmed and would allow the trp residues to be further used as intrinsic probes of LADH.

ACKNOWLEDGMENT

Unpublished work from this laboratory was supported by National Science Foundation grant DMB 88-06113.

REFERENCES

1. Abbreviations: d_T, dynamic depolarization factor, ranges from 0 for complete depolarization to 1.0 for no depolarization; IB, isobutyramide; Φ, fluorescence quantum yield; J, spectral overlap integral for resonance energy transfer; k_q^s, rate constant for solute quenching of an excited singlet state; k_q^T, rate constant for solute quenching of an excited triplet state; k_T, rate constant for resonance energy transfer; κ^2, orientation factor for resonance energy transfer; LADH, alcohol dehydrogenase from horse liver; NAD$^+$, oxidized β-nicotinamide adenine dinucleotide; NADH, reduced β-nicotinamide adenine dinucleotide; r, distance in angstroms between sites; r_o^{ss}, steady state fluorescence anisotropy; R_o, critical distance for resonance energy transfer at 50% efficiency, $R_o^{2/3}$, critical distance for energy transfer with κ^2 assumed to be $2/3$; τ^s, fluorescence lifetime of an excited singlet state; τ^T, phosphorescence lifetime of a triplet state; TFE, 2,2,2-trifluoroethanol.

2. Parker, C. A. Photoluminescence of Solutions; Elsevier: New York, 1968.

3. Chen, R. F.; Edelhoch, H., Eds. *Biochemical Fluorescence: Concepts*; Dekker: New York, 1975; Vols. 1 and 2.

4. Steiner, R. F., Ed. *Excited States of Biopolymers*; Plenum: New York, 1988.

5. Lakowicz, J. R. *Principles of Fluorescence Spectroscopy*; Plenum: New York, 1983.

6. Lakowicz, J. R., Ed. *Fluorescence Spectroscopy: Principles and Techniques*; Plenum: New York, in press.

7. Badea, M. G.; Brand, L. *Methods Enzymol.* **1979**, *61*, 378–425.

8. Demchenko, A. P. *Ultraviolet Spectroscopy of Proteins*; Springer-Verlag: New York, 1987.

9. Eklund, H.; Nordstrom, B. Zeppezauer, E.; Soderlund, G.; Ohlsson, I.; Boiwe, T.; Soderberg, B.-O.; Tapia, O.; Branden, C.-I. *J. Mol. Biol.* **1976**, *102*, 27–59.

10. Branden, C.-I.; Jornvall, H.; Eklund, H.; Furugren, B. (1975) *Enzymes*, 3rd. ed., *11*, 104–190.

11. Eklund, H.; Samama, J.-P.; Wallen, L.; Branden, C.-I.; Akeson, A.; Jones, T. A. *J. Mol. Biol.* **1981**, *146*, 561–587.

12. Eklund, H.; Samama, J.-P.; Jones, T. A. *Biochemistry* **1984**, *23*, 5982–5996.

13. Pettersson, G. *Critical Rev. Biochem.* **1987**, *21*, 349–389.

14. Brand, L.; Everse, J.; Kaplan, N. O. *Biochemistry* **1962**, *1*, 423–434.

15. Purkey, R. M.; Galley, W. C. *Biochemistry* **1970**, *9*, 3569–3575.

16. Laws, W. R.; Shore, J. D. *J. Biol. Chem.* **1978**, *253*, 8593–8597.

17. Abdallah, M. A.; Biellmann, J. F.; Wiget, P.; Joppich-Kuhn, R.; Luisi, P. L. *J. Biochem.* **1978**, *89*, 397–405.

18. Eftink, M. R.; Selvidge, L. A. *Biochemistry* **1982**, *21*, 117–125.

19. Eftink, M. R.; Jameson, D. M. *Biochemistry* **1982**, *21*, 4443–4449.

20. Hagaman, K. A.; Eftink, M. R. *Biophys. Chem.* **1984**, *20*, 201–207.

21. Luminescence studies with the LADH·NAD$^+$ binary complex are difficult since trace amounts of ethanol will cause NAD$^+$ to be catalytically converted to NADH.[39,97] For this reason, ternary complexes of NAD$^+$ with either pyrazole or TFE are usually studied.

22. Ross, J. B. A.; Schmidt, C. J.; Brand, L. *Biochemistry* **1981**, *20*, 4369–4377.

23. Knutson, J. R.; Walbridge, D. G.; Brand, L. *Biochemistry* **1982**, *21*, 4671–4679.

24. Eftink, M. R. Hagaman, K. A. *Biochemistry* **1986**, *25*, 6631–6637.

25. Demmer, D. R.; James, D. R.; Steer, R. P.; Verrall, R. E. *Photochem. Photobiol.* **1987**, *45*, 39–48.

26. Eftink, M. R. *Biophys. J.* **1987**, *51*, 278a.

27. Saviotti, M. L.; Galley, W. C. *Proc. Natl. Acad. Sci. USA* **1974**, *71*, 4154–4158.

28. Strambini, G. B.; Gonnelli, M. *Biochemistry* **1990**, *29*, 196–203.

29. Rousslang, K.; Allen, L.; Ross, J. B. A. *Photochem. Photobiol.* **1989**, *49*, 137–143.

30. Theorell, H.; McKinley-McKee, J. S. *Acta Chem. Scand.* **1961**, *15*, 1811–1833.

31. Gafni, A.; Brand, L. *Biochemistry* **1976**, *15*, 3165–3171.

32. Gafni, A., Schlessinger, J.; Steinberg, I. Z. *J. Am. Chem. Soc.* **1979**, *101*, 463–467.

33. Gafni, A. *Biochemistry* **1979**, *18*, 1540–1545.

34. Heitz, J. R.; Brand, L. *Arch. Biochem. Biophys.* **1971**, *144*, 286–291.

35. Conrad, R. H.; Heitz, J. R.; Brand, L. *Biochemistry* **1970**, *9*, 1540–1546.

36. Kovar, J.; Matyska, L.; Zeppezauer, M.; Marel, W. *Eur. J. Biochem.* **1986**, *155*, 391–396.

37. Wasylewski, Z.; Sucharski, P.; Wolak, A.; Eftink, M. R. *Biochim. Biophys. Acta* **1987**, *913*, 210–218.

38. While the data from most laboratories are consistent with the k_q^s for iodide quenching of Trp-314 to be zero,[16,17,18,25] Barboy and Feitelson[40] report an estimate of $1 \times 10^8 \, M^{-1} s^{-1}$ for this reaction.

39. Vekshin, N. L. *Eur. J. Biochem.* **1984**, *143*, 69–72.

40. Barboy, N.; Feitelson, J. *Photochem. Photobiol.* **1985**, *41*, 9–13.

41. Lakowicz, J. R.; Weber, G. *Biochemistry* **1973**, *12*, 4171–4179.

42. Lakowicz, J. R.; Maliwal, B. P.; Cherek, H.; Balter, A. *Biochemistry* **1983**, *22*, 1741–1752.

43. Eftink, M. R. In *Fluoresence Spectroscopy: Principles and Techniques*; Lakowicz, J. R., Ed.; Plenum: New York (in press).

44. Calhoun, D. B.; Vanderkooi, J. M.; Woodrow, G. V., III; Englander, S. W. *Biochemistry* **1983**, *22*, 1526–1532.

45. Eftink, M. R.; Wasylewski, Z.; Ghiron, C. A. *Biochemistry* **1987**, *26*, 8338–8346.

46. Gafni, A. *J. Am. Chem. Soc.* **1980**, *102*, 7367–7368.

47. Kishner, S.; Trepman, E.; Gallay, W. C. *Can. J. Biochem.* **1979**, *57*, 1299–1304.

48. Strambini, G. B. *Biophys. J.* **1987**, *52*, 23–28.

49. Gabellieri, E.; Strambini, G. B.; Gualtieri, P. *Biophys. Chem.* **1988**, *30*, 61–67.

50. Calhoun, D. B.; Vanderkooi, J. M.; Englander, S. W. *Biochemistry* **1983**, *22*, 1533–1539.

51. Calhoun, D. B.; Englander, S. W.; Wright, W. W.; Vanderkooi, J. M. *Biochemistry* **1988**, *27*, 8466–8474.

52. Vanderkooi, J. M.; Englander, S. W.; Papp, S.; Wright, W. W.; Owen, C. S. *Proc. Natl. Acad. Sci. USA* **1990**, *87*, 5099–5103.

53. Gafni, A.; Mesol, J. V.; Steel, D. G. In *Time-Resolved Laser Spectroscopy in Biochemistry II*; Lakowicz, J. R., Ed., Proc. SPIE; Bellingham, WA, 1990; vol. 1204; pp. 763–764.

54. Theorell, H.; Tatemoto, K. *Arch. Biochem. Biophys.* **1971**, *142*, 69–82.

55. Holbrook, J. J.; Yates, D. W.; Reynolds, S. J.; Evans, R. W.; Greenwood, C.; Gore, M. G. *Biochem. J.* **1972**, *128*, 933–940.

56. Lakowicz, J. R.; Balter, A. *Biophys. Chem.* **1982**, *16*, 99–115 and 117–132.

57. Beechem, J. R. M.; Gratton, E. In *Time-Resolved Laser Spectroscopy in Biochemistry II*; SPIE Proc. Bellingham, WA, 1988; Vol. 909; pp. 70–81.

58. Yamamoto, U.; Tanaka, J. *Bull. Chem. Soc. Japan* **1972**, *45*, 1362–1366.

59. Eftink, M. R.; Selvidge, L. A.; Callis, P. R.; Rehms, A. A. *J. Phys. Chem.* **1990**, *94*, 3469–3479.

60. Shore, J. D., Gutfreund, H.; Yates, D. *J. Biol Chem.* **1975**, *250*, 5276–5277.

61. Subramanian, S.; Ross, J. B. A.; Ross, P. D.; Brand, L. *Biochemistry* **1981**, *20*, 4086–4093.

62. Holmes, L. G.; Robbins, F. M. *Photochem. Photobiol.* **1974**, *19*, 361–366.

63. Weers, J. G.; Maki, A. H. *Biochemistry* **1986**, *25*, 2897–2904.

64. Brand, L.; Knutson, J. R.; Davenport, L.; Beechem, J. M.; Dale, R. R.; Walbridge, D. G.; Kowalczyk, A. A. In *Spectroscopy and the Dynamics of Molecular Biological Systems*; Bayley, P. M.; Dale, R. E., Eds.; Academic: London, 1985; 259–305.

65. Ghiron, C. A.; Longworth, J. W. *Biochemistry* **1979**, *18*, 3828–3832.

66. Desie, G.; Boens, N.; DeSchryver, F. C. *Biochemistry* **1986**, *25*, 8301–8308.

67. Eisinger, J.; Feuer, B.; Lamola, A. A. *Biochemistry* **1969**, *8*, 3908–3915.

68. Weber, G.; Shinitzky, M. *Proc. Natl. Acad. Sci. USA*, **1970**, *65*, 823–830.

69. Bucci, E.; Steiner, R. F. *Biophys. Chem.* **1988**, *30*, 199–224.

70. Lakowicz, J. R.; Laczko, G.; Gryczynski, I.; Cherek, H. *J. Biol. Chem.* **1986**, *261*, 2240–2245.

71. Eftink, M. R. *Biophys. J.* **1983**, *43*, 323–333.

72. Strambini, G. B.; Gabellieri, E. *Biochemistry* **1987**, *26*, 6527–6530.

73. Strambini, G. B.; Gonnelli, M. *Chem. Phys. Lett.* **1985**, *115*, 196–200.

74. Papp, S.; Vanderkooi, J. M. *Photochem. Photobiol.* **1989**, *49*, 775–784.

75. Theorell, H.; Tatemoto, K. *Arch. Biochem. Biophys.* **1971**, *142*, 69–82.

76. Anderson, D. C.; Dahlquist, F. W. *Biochemistry* **1982**, *21*, 3569–3578.

77. Eftink, M. R.; Hu, J. submitted for publication in *Biochemistry*.

78. Theorell, H.; Yonetani, T. *Biochem. Z.* **1963**, *338*, 537–553.

79. Shore, J. D. *Biochemistry*, **1969**, *8*, 1588–1590.

80. Bernard, S. A.; Dunn, M. F.; Luisi, P. L.; Shack, P. *Biochemistry* **1970**, *9*, 185–192.

81. Dunn, M. F.; Bernard, S. A.; Anderson, D.; Copeland, A.; Morris, R. G.; Rogue, J. P. *Biochemistry* **1979**, *18*, 2346–2354.

82. Sekhar, V. C.; Plapp, B. V. *Biochemistry* **1988**, *27*, 5082–5088; Sekhar, V. C.; Plapp, B. V. *Biochemistry* **1990**, *29*, 4289–4295.

83. DeTraglia, M. C.; Schmidt, J.; Dunn, M. F.; McFarland, J. T. *J. Biol. Chem.* **1977**, *252*, 3493–3500.

84. Strambini, G. B.; Gonnelli, M. *Biochemistry* **1990**, *29*, 203–208.

85. Parker, D. M.; Hardman, M. J.; Plapp, B. V.; Holbrook, J. J.; Shore, J. D. *Biophys. J.* **1978**, *173*, 269–275.

85a. Walbridge, D. G.; Knutson, J. R.; Han, M. K.; Brand, L. *Biophys. J.* **1987**, *51*, 284a.

86. Wolfe, J. K.; Weidig, C. F.; Halvorson, H. R.; Shore, J. D.; Parker, O. M.; Holbrook, J. J. *J. Biol. Chem.* **1977**, *252*, 433–436.

87. Eftink, M. R. *Biochemistry* **1986**, *25*, 6620–6624.

88. Laws, W. R.; Shore, J. D. *J. Biol. Chem.* **1979**, *254*, 2582–2584.

89. Gonelli, M.; Strambini, G. B. *Biophys. Chem.* **1986**, *24*, 161–167.

90. Strambini, G. B.; Gonnelli, M. *Biochemistry* **1986**, *25*, 2471–2476.

91. Vallee, B. L.; Hoch, F. L. *J. Biol. Chem.* **1957**, *225*, 185–195.

92. Dietrich, H.; MacGibbon, A. K. H.; Dunn, M. F.; Zeppezauer, M. *Biochemistry* **1983**, *22*, 3432–3438.

93. Baumgartem B.; Hones, J. *Photochem. Photobiol.* **1988**, *47*, 201–205.

94. Sun, M.; Song, P. S. *Photochem. Photobiol.* **1977**, *25*, 3–9.

95. Valeur, B.; Weber, G. *Photochem. Photobiol.* **1977**, *25*, 441–444.

96. Evleth, E. M. *J. Am. Chem. Soc.* **1967**, *89*, 6445–6453.

97. Tanigushi, S. *Acta Chem. Scand.* **1967**, *21*, 1511–1518.

98. Barboy, M.; Feitelson, J. *Biochemistry* **1978**, *23*, 4923–4926.

99. Walbridge, D. G.; Knutson, J. R.; Brand, L. *Anal. Biochem.* **1987**, *161*, 467–478.

100. Han, M. R.; Walbridge, D.; Knutson, J. R.; Neyroz, P.; Brand, L. In *Fluoresence Biomolecules*; Jameson, D. M.; Reinhart, G. D., Eds.; Plenum: New York, 1989, 33–59.

101. Malin, E. L.; Young, J. M. *Alcohol Aldehyde Metab. Syst.* **1977**, *2*, 137–143.

102. Beechem, J. M.; Knutson, J. R.; Ross, J. B. A.; Turner, B. V.; Brand, L. *Biochemistry* **1988**, *22*, 6054–6058.

103. Coates, J. H.; Hardman, M. J.; Shore, J. D.; Gutfreund, H. *FEBS Lett.* **1977**, *84*, 25–28.

104. Beechem, J. M.; Knutson, J. R.; Brand, L. *Trends. Biochem. Soc.* **1987**, *14*, 832–835.

105. Beechem, J. R. *Biophys. J.* **1990**, *57*, 430a.

106. Sprangler, C. J.; Brand, L. *Biophys. J.* **1991**, *59*, 359a.

107. Strambini, G. B. *Biophys. J.* **1988**, *43*, 127–130.

SURFACE-ENHANCED RESONANCE RAMAN SCATTERING (SERRS) SPECTROSCOPY:
A PROBE OF BIOMOLECULAR STRUCTURE AND BONDING AT SURFACES

Therese M. Cotton, Jae-Ho Kim, and Randall E. Holt

Advances in Biophysical Chemistry, Volume 2, pages 115–147

ISBN: 1-55938-396-8

I. INTRODUCTION

The technique of surface-enhanced resonance Raman scattering (SERRS) spectroscopy uniquely combines the advantages of resonance Raman and surface-enhanced Raman scattering (SERS) spectroscopy. Advantages of resonance Raman spectroscopy include enhancement in the scattering cross-section due to involvement of the electronic transition dipole moment in the scattering process. The large polarizability change that occurs within a molecule when it undergoes an electronic transition results in an enhancement of the Raman scattering intensity by as much as 10^6-fold over the non-resonant Raman process, allowing analysis of low concentrations of material, typically as low as 10^{-6} M. For chromophores with high symmetry resonance Raman spectroscopy may provide information about the symmetries of the electronic states involved in the transition. The enhancement of the chromophore vibrational modes make resonance Raman attractive as a method of selective analysis for the chromophore within a non-chromophoric matrix, such as protein. Finally, the small scattering cross-section of water allows the observation of a Raman spectrum in aqueous solutions, the usual media for biological preparations. These advantages have made resonance Raman spectroscopy invaluable for the study of many biological systems that contain chromophores, including heme-containing proteins, photosynthetic systems, visual pigments, carotenoids and flavins. Comparison of the spectra of free chromophores to those observed from chromophores bound in a biological matrix provides valuable information about the environment and nature of the bonding interactions.

For some applications, resonance Raman spectroscopy may not be sufficiently sensitive to analyze limited amounts of valuable material. For low volumes of material at low concentrations, or for particularly labile material, preconcentration may not be possible. Furthermore, intense chromophore fluorescence may mask the Raman scattering. Because of the usefulness of the resonance Raman measurement, considerable effort has been expended toward devising techniques to circumvent fluorescence interference. For example, judicious choice of the excitation and detection wavelengths may allow observation of a fluorescence-free resonance Raman spectrum. Other strategies include addition of a fluorescence quencher, the use of instrumentally sophisticated techniques such as coherent anti-Stokes Raman scattering (CARS)[1] spectroscopy, inverse Raman spectroscopy,[1] Raman gain spectroscopy[1] and temporal[2] and phase[3] resolved Raman scattering.

When molecules are adsorbed on or near metal surfaces, their electronic properties may be modified in two ways: by direct interaction of molecular orbitals with those of the metal, resulting in chemisorption or, for weaker interactions, physisorption; and by modification of the electric field within and around the molecule. The interaction of adsorbates with certain metals results in a large enhancement of the Raman scattering cross-section, comparable to that observed in resonance Raman scattering, typically 10^6-fold. This SERS process appears to be characteristic of the metal rather than the adsorbate, based upon the wide range of adsorbates that have been reported to experience surface-enhancement. Molecules that undergo both resonance and surface-enhancement may experience a total enhancement as high as 10^{12}-fold over normal Raman scattering. Furthermore, fluorescence is often quenched near metal surfaces. The enhanced Raman scattering and fluorescence quenching predicted for combined resonance and SERS make the application of this technique particularly attractive for the study of biological materials.

In this chapter, the current theories that have been proposed to account for the SERS effect, particularly as applied to the resonance SERS (or SERRS) experiment will be examined first. Next, a review of the SERRS literature, pertaining the application of SERRS to the study of dyes, and then to the study of biological systems will be provided. In this review, the scope of applications of the technique, its limitations and its strengths, as well as the degree to which current theoretical treatments account for the manifestations of the effect will be covered.

II. SERS/SERRS THEORIES

A. Experimental Observations of the SERS/SERRS Effect

Since the first reports of enhancement of the Raman scattering cross-section at an electrochemically roughened Ag electrode[4] in 1977 by Jeanmaire and Van Duyne[5A] and by Albrecht et al.,[5B] enhanced Raman scattering of many different compounds has been reported and the list of metals reported to give rise to the enhancement effect has increased. A recent bibliography[6] lists over 130 different compounds or classes of compounds reported to undergo enhancement at the silver electrode in an electrochemical system. This number is even greater when compounds reported to experience enhancement at other silver substrates (colloidal suspensions, solid/air interfaces, solid/vacuum (or gas) interfaces, silver/metal (or nonmetal) matrices) are included. Representative classes of these compounds include inorganic and organic cations and anions, neutral gases, metal complexes, dyes, molecules of biological importance (amino acids, proteins, nucleic acids), and more highly organized biological assemblies such as photosynthetic membranes and reaction centers. In addition to Ag, there are other metals that produce SERS enhancement: Au, Cu, Ni, Pt, Pd, Rh, Ti, Li, Na, K, Cd, In and Co. The metal oxides TiO_2 and NiO have been reported to produce surface enhancement. This list is based upon literature reports with no effort made to substantiate the observations.

Although the giant enhancement of Raman scattering is the most characteristic manifestation of the SERS effect, a number of other effects are commonly observed. Efrima[7] lists the following criteria (with associated references) that can be used to distinguish a surface spectrum from a bulk signal.

1. Band frequency shifts.
2. Changes in relative band intensities that may vary with potential and excitation wavelength. Such changes may also occur for the same molecule on different SERS active substrates.
3. Appearance of new bands in the spectrum. These may be vibrational bands of the adsorbed molecule. These bands are weak or forbidden in the bulk sample or may result from chemical changes in the adsorbed molecule.
4. Differences between the surface and solution Raman excitation profiles.
5. Depolarization of all bands in the spectrum.
6. Greater scattering of p-polarized light (incident electric field polarized perpendicular to the plane of the surface) compared to s-polarization.
7. High continuous background associated with the scattering process.
8. Dependence of the signal on electric potential in electrochemical systems. This may be the result of:
   a. electric-potential dependent adsorption and/or desorption;
   b. morphology changes of the electrode surface;
   c. an intrinsic dependence of the scattering process itself on the electric potential;
   d. electrogeneration of a new species;
   e. electric potential dependent reorientation.
9. Dependence of the surface signal on concentration in a manner different from that of the bulk signal.
10. Changes in the surface scattering from an adsorbed molecule by the presence of other chemical species differ from that of the same molecule in solution.
11. Fluorescence quenching in dye molecules.

A more recent review of SERS at electrodes by Chang[8] adds the following metals to the list of those producing SERS enhancement: Al, Fe and Sn. In addition to H_2O as the usual electrochemical solvent, Chang lists references for the use of the non-aqueous solvents acetonitrile, N,N'-dimethylformamide and propylene carbonate in SERS studies. Other observations associated with SERS, such as experimental modifications reported to produce further increases or decreases in SERS, are also described in this review.

An extensive amount of theoretical work has been reported in an attempt to account first for the large surface enhancement and, to variable degrees, many of the other characteristics of the SERS effect, as described above. An adequate theoretical model should account for these characteristics, the wide range of molecules reported to undergo surface-enhancement, the different metals (particularly Ag, Au and Cu) reported to produce the effect, and other experimental observations, such as the dependence of signal intensity upon surface–molecule distance and the dielectric constant of the surrounding medium. Generally, the theories proposed may be classified into two groups: those involving effects of the surface upon the local and scattered electromagnetic fields ("electromagnetic" or EM enhancement) and those involving a chemical interaction between the molecule and the surface resulting in modification of the molecular polarizability ("chemical" enhancement). The distinction between these two classes is not well-defined, however, because electromagnetic enhancement may effect the molecular polarizability. The classification will nevertheless be used and distinction made according to the dominant factors operative in the model. Excluded from the consideration of theory is the lesser well-accepted theory of SERS enhancement due to the presence of carbon at the metal surface.[9–11]

In the specific instance where the adsorbed molecule is excited to a higher electronic state in the SERS experiment, the excited molecule may also interact with the surface. When this occurs, the resulting SERRS spectrum is dependent upon the lifetime of the excited state and the polarizability along the transition dipole moment. An adequate model must also account for the manifestation of these effects upon the SERRS enhancement factor and the fluorescence yield.

B. EM Enhancement Theories

A number of excellent reviews of EM processes are available.[7,12–15] Wokaun[12] provides a comprehensive review of those models that may be classified as EM models but which differ in their mechanisms and predictions. Although each of the models proposed contains elements that are supported by experiment, the greatest body of experimental evidence appears to support the involvement of the localized particle plasmon model.[7,12]

When the adsorbed species possesses an electronic resonance that can be excited in the SERS experiment, modifications to the EM theory must be made. Consideration must be given to the interaction of the oscillating electronic transition dipole moment with the metal surface, which may affect the optical properties of the adsorbed species (e.g., absorption and fluorescence scattering) as well as the resonance Raman scattering intensity. Accordingly, a body of theoretical work has been developed to treat SERRS enhancement of chromophores at metal surfaces.[16–22]

The electrodynamic effects imposed on a scattering molecule by a nearby metal surface has been considered in detail by Metiu.[13] Using a model consisting of a molecular dipole located near a metal surface, he calculated the effect of the metal surface on the local and scattered EM fields. The dependence of the enhancement,

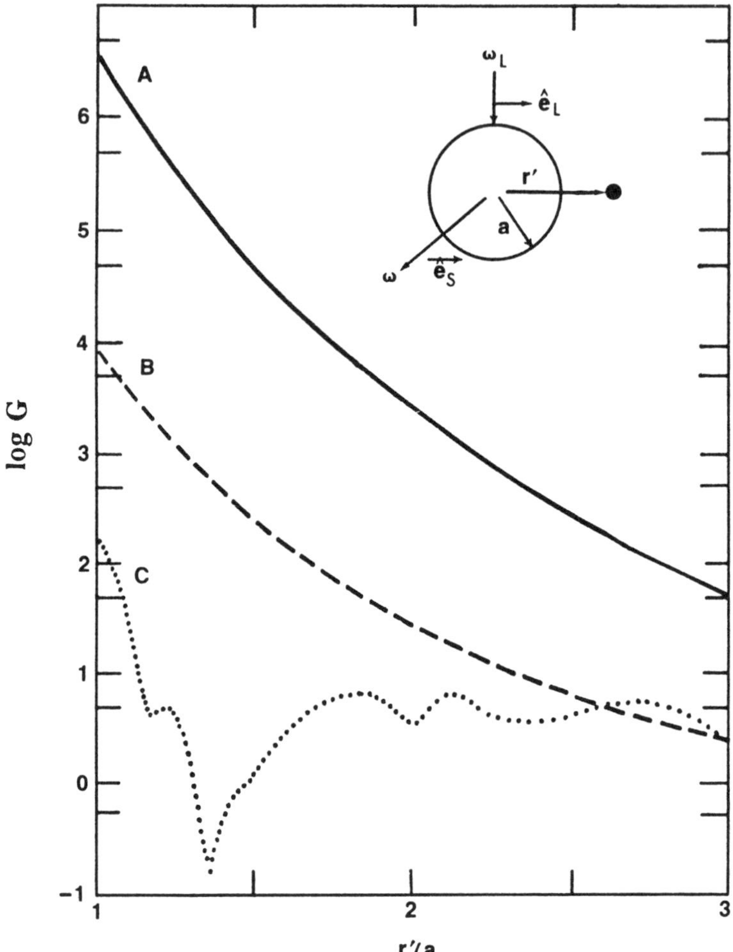

Figure 1. The relationship between the enhancement (G) and the distance (r') of a molecule (solid circle) from the center of the silver sphere for three different radii (a); A: $a = 5$ nm and excitation wavelength (λ_0) = 382 nm; B: $a = 50$ nm and (λ_0) = 500 nm; C: $a = 500$ nm and (λ_0) = 528 nm. e_L and e_S are the directions of the incident and scattered field. ω_L is the frequency of the primary exciting field.[15]

G, on the distance of the molecule from the center of the sphere, r, for three different sphere radii, a, and three different excitation wavelengths λ_0 is shown in Figure 1. The enhancement reaches a maximum value when the dielectric constant of the surrounding medium is equal to 1 and depends on (a/r').[12] This relationship reflects the fourth power of the inverse cube dependence of the dipole field at the substrate.[15]

The effects of the surface upon the EM fields in the system include:

1. an increase in local (with respect to the dipole) EM field due to a reflected component of the incident EM field;
2. modification of the local EM field due to scattering by an induced image dipole; and
3. an increase in scattered EM field due to scattering by induced metal resonances.

In the cases of Rayleigh and normal Raman scattering, the response of the molecule to the local EM field is described by the Drude-Lorentz treatment,[23] which adequately describes the properties of a scattering system for most experiments.[24–26] The effects of the enhanced local EM field on the molecular properties are predicted to be:

1. an increase in the transition oscillator strength (absorption coefficient) due to the reflected field;
2. an increase in excited state bandwidth; and
3. a small shift in the oscillating dipole frequency.

Details of the calculations and estimates of the enhancement factors for Rayleigh, Raman and fluorescence scattering are provided by Metiu.[13] An important result of these calculations is that the enhancement factor for resonance Raman scattering is predicted to equal the square root of the enhancement factor for normal Raman scattering.

Calculations of scattering properties at other types of surfaces are also summarized. Predictions of scattering at smooth metal surfaces[27,28] indicate that enhancement of the local EM field is described by the Fresnel equation[29] and a normal Raman scattering enhancement of ca. 30 for a system of molecular dipoles oriented perpendicular to the metal surface. Only bulk metal electron resonances are involved via an oscillating image dipole that adds to the total scattered field.

If an electromagnetic resonance in the metal can be directly excited by the incident electric field, the local electric field would be increased substantially. Such electromagnetic resonances in metals may be sustained by *plasmons*, or quanta associated with electron oscillations within the metal.[30]

The lowest order particle plasmon is excited into resonance at frequencies typically employed in the Raman experiment. This particular resonance is constrained to the surface of the metal and is identified as the *localized surface plasmon*.[31] A physical picture of the surface plasmon is an electric field wave that moves along the surface of the metal.

Under usual conditions, the excitation of a surface plasmon by an incident photon is prohibited by conservation of momentum considerations.[13] Momentum match can be produced by using defined experimental arrangements: coupling of the incident radiation to the surface in an attenuated total reflectance configuration, modifying the surface to act as a grating, and producing small scale (<1000 Å) roughness on the surface. Similarly, metal spheres and spheroids of appropriate dimensions will couple incident radiation with surface plasmon resonances. The wavelength dependence of surface plasmon excitation is a function of the dielectric properties of the metal. Based upon this dependence,[31] (the strongest resonances in the visible region are induced in Ag, Au and Cu, those metals that are observed to produce the best SERS enhancement.[7] Extension of the theory to large scale[32] (dimensions larger than $\frac{1}{15}$ of the excitation wavelength) and small scale (<100 Å) roughness has been reviewed.[8,15] The effects of the roughness geometry on the surface plasmon resonance[33] and particle dipolar interactions was also established.[12]

Major predictions of the local plasmon model (for small spheres) include:

1. local intensity enhancements approaching 10^{11} (for normal Raman, or SERS), with average enhancements near 10^5–10^6;
2. a relationship between enhancement and distance proportional to $(r/a)^{12}$; r = metal–molecule separation, a = local radius of curvature of metal surface;[34]
3. greatest enhancement from Ag, with lesser enhancements from Cu and Au (at lower excitation frequencies), and even lower enhancements from other metals;
4. limited molecular specificity based upon the differing polarizabilities of various compounds;
5. depolarization of the Raman spectrum due to the sizable contribution of the plasmon resonances to the total emission;
6. normal (non-resonance) Raman excitation profile approximating the plasmon absorbance. Maximum enhancement is expected at excitation wavelengths near the maximal absorbance in the visible region of the spectrum in the case of Ag films and colloids or near the region of minimal reflectance in the case of Ag electrodes.

The treatment described by Metiu[13] provides a phenomenological picture of the influence of the metal surface upon the properties of an adsorbed scatterer. The simple two-level model implicit in the Drude–Lorentz polarizability theory used in this treatment is inadequate for describing the response of a real molecular dipole when an electronic resonance is excited.

Weitz et al.[20] describe a four-level model, two vibronic levels in the ground state and two in the excited electronic state. Using this model, the authors describe the effects of the surface upon normal Raman, resonance Raman and fluorescence scattering. According to this model, the effects of the surface are threefold: (1) enhancement of the local EM field, which produces an enhanced absorption by the molecular dipole, (2) enhancement of the scattered EM field, and (3) the availability of an additional non-radiative mechanism for decay of the molecular excited state. This additional decay channel decreases the emission yield and lifetime of the excited state. It also reduces the magnitude of the resonance Raman scattering enhancement by as much as 10^3 relative to the enhancement of normal Raman scattering. The effect of this additional decay mechanism upon fluorescence depends upon the quantum efficiency (QE) of the fluorophore. For species having a high fluorescence QE, fluorescence is quenched at the surface. In contrast, those species that have a low QE will experience fluorescence enhancement. These predictions are verified for scattering from *p*-nitrobenzoate, rhodamine 6-G and basic fuchsin on Ag island films, results that will be reviewed in Section 1.3.2.5.

A modification of the four-level model was presented by Efrima.[21] This model includes many excited state rovibronic levels in the scattering process. The effects of this modification are to increase the predicted resonance Raman enhancement factor so that the enhancement of normal Raman and resonance Raman scattering are nearly equal. The fluorescence enhancement factor is predicted to be much smaller in this modification.

Absolute enhancement factors cannot be calculated for either of the models described above. Absolute enhancement factors were obtained by Zeman et al.[22] using electrodynamic calculations. This method is prevalent in theoretical treatments describing SERS[33] and early treatments of SERRS.[13,14] The calculations were performed to account specifically for the SERRS scattering of cobalt phthalocyanine on CaF_2 roughened glass surfaces coated with a thick (1000 Å) Ag film. The enhancement factor was found to maximize at low (0.1 ML) coverages. Greater coverages produce significant fall-off in enhancement. Calculations also suggest that intermolecular dipole–dipole coupling between neighboring adsorbates (on the Ag surface) has only a small influence on the SERRS enhancement factor. The wavelength dependence of the enhancement reflects the involvement of both the metal and molecular resonances.

Many of the experimental manifestations of SERS can be successfully accounted for by EM enhancement theories. Surface selection rules that describe the effect of adsorbate orientation on SERS band intensities have been developed[35–37] and reviewed.[36] The rules are validated by results on Ag films[37] and Ag electrodes[38,39] for the non-resonance case where enhancement is dictated by the metal plasmon resonances only. When molecular resonance controls the band intensities, the selection rules should be modified. Enhancement of the molecular resonance process[13,20] will produce enhancement of all modes that experience enhancement in the absence of the surface. Hence, for the non-resonance case, major differences in relative band intensities may result from different orientations of the adsorbate at the surface, whereas for the resonance case, only differences in overall spectral intensity should be observed. The distance dependence of SERS[34,40,41] and the appearance of Raman "forbidden" bands in the SERS spectrum are also successfully accounted for by EM considerations.

C. Chemical Enhancement Theories

The development of non-electromagnetic or "chemical" enhancement theories has paralleled that of EM theory. Early predictions of enhancement factors as large as 10^6 based upon chemical models sparked considerable interest in the pursuit of a chemical explanation of the surface enhancement effect. As the body of experimental results accumulated, it became clear that other manifestations must be accounted for besides the enhancement of the scattering cross-section. Although EM enhancement theory can adequately account for most of the observations, it fails to explain others. Thus, in spite of the success of EM enhancement theory, chemical enhancement theories are still the subject of significant effort.

Chemical enhancement models require interaction of the adsorbed molecule with the metal surface in an intimate fashion, resulting in the mixing of molecule and metal electronic states. Three types of interaction may be envisioned. The first involves chemisorption of the molecule onto the metal and overlap of the molecule and metal orbitals. This surface complex possesses physical properties that are different from those of the free molecule. If the electronic energies of the complex are sufficiently shifted from those of the free molecule, molecular resonance may be induced at excitation wavelengths far removed from that of the free molecule. This type of enhancement may be considered as surface-induced resonance enhancement, but it is produced by a complex that is chemically distinct from the free molecule. A second type of interaction involves the donation of electron density via a donor-acceptor complex. In this interaction, the energies of the electronic states are not significantly perturbed, but the polarizability of either the molecule or metal is increased by electron donation. The interaction is mediated either through covalent bonding or through the formation of a ground-state charge transfer (CT) complex. This type of interaction may be considered to be surface-induced hyperpolarizability. The third type of interaction is described by an increase in polarizability of either the molecule or metal through an excited state CT interaction, termed photon-driven charge transfer. In this case, the ground state polarizabilities are not significantly perturbed but excitation by a photon allows a transition to occur from an occupied orbital of the molecule to the metal or vice versa.

All chemical enhancement theories focus on the surface induced changes in the polarizability of the metal-molecule system. This is in contrast to the EM enhancement theories which focus on enhancement of the local electromagnetic fields. Each of the three types of chemical enhancement models described above will be considered in detail.

1. Surface-induced Resonance Enhancement

This model was one of the earliest proposed to account for the surface enhancement of pyridine on Ag.[42] Differential reflectance spectra of a single crystal

Ag (111) electrode showed that during anodization a complex was formed between Ag-pyridine and halide ions present in the electrolyte, by the appearance of a broad absorption band near 750 nm. Electroreflectance spectra of the same surface revealed that the anodization conditions which were used did not produce surface roughness in the presence of the halide ions, thus excluding a surface plasmon mechanism. The presence of the new absorption band at 750 nm supports a mechanism whereby scattering from the complex is enhanced through a conventional resonance Raman process. An enhancement factor of 10^6 was observed and attributed to this process, in agreement with typical enhancement factors resulting from a pure resonance Raman process.

Predictions made by this model are intuitively obvious. The new electronic states of the complex should produce a Raman excitation profile that is greatly different from that of the solution species. Enhancement is only expected to occur for the molecules in direct contact with the metal surface; thus enhancement is short range. The enhancement is expected to be both metal and molecule specific, limited to those compounds that are capable of forming a specific complex. Potential dependence may result from the potential dependent behavior of the components of the complex (i.e., adsorption/desorption of ions, molecule).

2. Surface-induced Ground State Hyperpolarizability

These models produce enhancement through the increase in electron density in either the adsorbed molecule or the metal as a result of the adsorption process, usually through a ground state charge-transfer interaction. Vibrations of the adsorbed species will modulate the charge-transfer process. Aussenegg and Lippitsch[43] calculated enhancement factors as great as 10^5–10^6 for pyridine on Ag.

A similar model was proposed by Abe et al.[44] to account for the enhancement of Raman scattering from chemisorbed CO on Ag colloids. Strong modes in the SERS spectrum were attributed to CO stretching, Ag-C stretching, and AgCO bending. The CO stretching excitation spectrum was observed to follow closely the absorption spectrum of the Ag colloidal suspension. This was interpreted to result from charge-injection into, and out of, the conduction-electron metal plasma with the CO stretching and subsequent scattering from the metal plasma.

The predictions made by these models depend upon the individual models. Enhancement factors near 10^6 are predicted. No surface roughness is required in these models. The Raman excitation profile should reflect the presence of a ground state charge-transfer complex by the appearance of a new absorption band associated with the CT process. Potential dependence may be reflected in the increase or decrease in intensity of the spectrum with a change in the metal charge. Modes that change the metal–molecule distance or modulate the charge-injection process may be preferentially enhanced depending upon the specific model. Only short-range enhancement is predicted. Metal and molecule specificity is predicted as is the case for the surface-induced resonance enhancement models.

3. Photon-driven Charge Transfer Models

The photoinduced charge-transfer (PDCT) mechanism may be described as a resonance Raman effect involving electronic states of both the adsorbed molecule and the metal. This mechanism was first proposed by Gersten et al.[45] and employed a two-state level with energies above and below the Fermi level of the metal. Real transitions between occupied and unoccupied levels in the system and transitions to bound virtual states cause a large enhancement of the scattering cross section of the adsorbate.

A similar model was proposed by Persson.[46] In this model the energy of an unoccupied level of an adsorbate is shifted and broadened upon adsorption due to Newns–Anderson resonances (short time scale excursions of metal electrons into an unoccupied orbital of the adsorbate). These resonances produce an effective partial filling of the adsorbate orbital. Absorption of a photon results in the transition of an electron from below the Fermi level to the maximum density of the adsorbate orbital. As with all of the CT models, if electron-hole pair (e–h pair) recombination results in an excited vibrational state of the adsorbate, the scattered photon will be Stokes shifted. The model predicts an enhancement of ca. 30.

A number of other PDCT models have been proposed involving similar transitions between states in the metal and adsorbate. Otto[47] proposed a four step charge-transfer model. An incident photon induces a transition within the metal to a virtual state (step 1). This is followed by electron transfer to an excited state of the adsorbate via tunneling (for physisorbed species) or hybridization (for chemisorbed species) (step 2). Electron transfer back to the metal occurs leaving the adsorbate in an excited vibrational ground state (step 3) followed by electron–hole recombination emitting a photon at the Stokes shifted frequency (step 4). An analogous process may occur involving electron transfer from the adsorbate to a hole created by absorption of a photon by the metal. The process is resonant when the energy difference of the metal and adsorbate levels of the first two steps equals the energy of the incident photon.

The excitation of electron–hole pairs is a requirement for the PDCT mechanism. The excitation of electron-hole pairs on a smooth surface is forbidden by inability to match momenta of the incident/scattered photons and the electron-hole pair.[8] When a molecule is located at the atomic scale clusters (adatom) of the noble metal, the induced dipole of the molecule has wave-vector uncertainty (ΔK) as shown on the right side of Figure 2. The left side shows the transitions that occur under normal conditions on a smooth surface. Thus, ΔK induces an enhancement of the probability for intraband transitions within the sp-band. The presence of roughness on the surface relaxes this conservation rule as in the plasmon model in EM theory. The scale of roughness, however, is not restricted by such considerations as the electron mean free path (as in EM theory). "Atomic" scale roughness was proposed based upon the apparent enhancement of Raman scattering on "smooth" Ag surfaces.[15] This proposition introduces the concept of SERS "active sites". Surface

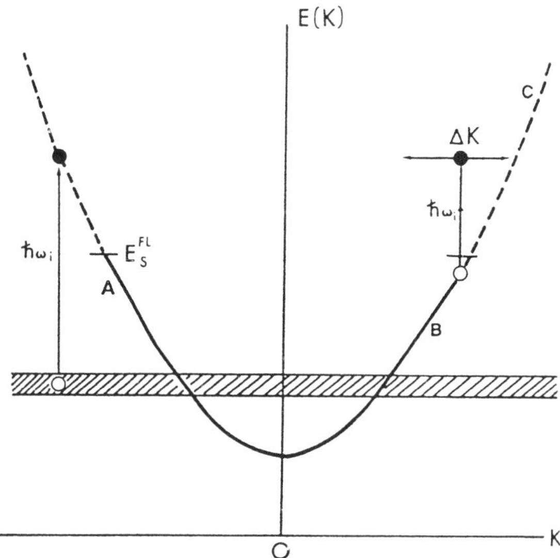

Figure 2. Schematic of the intraband transitions of the adsorbate on the adatom (right side of the diagram). An electron is ejected from the filled sp-band (open circle) by a photon and enters the empty sp-band (Solid circle). A: Fermi level; B: filled sp-band; C: empty sp-band; E_d: energy level of d-band.[8]

defects such as bulk defects, and vacancies in single crystals and adatoms (single metal atoms located at the surface of the crystal plane)[48] are sufficient to allow coupling of photons to e–h pair formation. Furthermore, SERS active sites may be the location of specific adsorption. Ag^+, "ad-ion" was proposed as an active site.[49] The Ag^+ induces the formation of a complex involving the adsorbate and electrolyte anions. The enhancement mechanism is suggested to be either surface-induced resonance or PDCT. The involvement of Ag_4^{+n} (n = 1 to 3) clusters was recently proposed.[50,51] The involvement of clusters accounts for features routinely observed in the low frequency SERS spectra of adsorbates on electrodes.[52] The existence of active sites accounts for the involvement of anions in the enhancement process via stabilization of the positively charged ad-ion or cluster.[49] The number of SERS active sites may be relatively small. Otto[48] estimates that 10% of the surface of thick (2000 Å), cold deposited Ag films consists of active sites. The overall enhancement is diminished with respect to the large enhancement at these sites by the averaging of the signal over the entire irradiated surface.

A different approach to the theoretical treatment of PDCT was taken by Adrian,[53] Lippitsch[54] and Lombardi et al.[55] The model involves transitions from a two-state adsorbate either to, or from, the metal Fermi level. The intensities of Raman scattered radiation from particular vibrational modes of the adsorbate are treated as resonance Raman scattering according to Tang et al.[56] The success of this treatment is shown by its ability to account accurately for the potential dependence of SERS intensity in certain electrochemical systems.[57]

The chemical enhancement theories described above also account for other experimental features of SERS/SERRS. Different orientations of an adsorbate at the metal surface are reflected in SERS spectra as differences in band frequencies.[58–61] The dependence on incident wavelength and applied potential have already been discussed.

D.   Chemical versus EM Enhancement Mechanisms

Determination of the enhancement mechanism in a SERS/SERRS scattering system is important. Chemical enhancement requires direct contact of the adsorbate with the metal surface. Electromagnetic enhancement can produce long range enhancement. (e.g., Murray[34] demonstrated using a polymer spacer on CaF_2 roughened Ag film assemblies that the surface-enhancement of p-nitrobenzoic acid decreased by only ca. 10-fold with a 50 Å metal—molecule distance). A chemical enhancement mechanism implies by its very nature a change in the chemical properties of the adsorbate, whereas this is not required in an EM mechanism. The predictions of the enhancement mechanisms with respect to SERS studies of biological materials are important. To maintain functional activity, the biological system must not be significantly perturbed. Additionally, the chromophores associated with most biological systems are located within a matrix (e.g., protein, membrane). The determination of the enhancement mechanism can aid in the assessment of the functional state of the SERRS scattering species. Evidence for chemical enhancement of a chromophore known to be buried within a protein would suggest that the chromophore has been removed from within the protein and is in direct contact with the metal surface. Additionally, determination of the enhancement mechanism can provide insight into the nature of the interaction of an adsorbate with the metal surface. This knowledge can provide insight into the dynamics of electron-transfer at electrode surfaces, for example. Accordingly, most SERS reports include some assessment of the enhancement mechanism involved in the respective systems.

In many cases, the exact nature of the enhancement process is difficult to determine. Many of the characteristics of the SERS effect can be accounted for by either EM or chemical enhancement theories. Some observations provide evidence for the predominance of one mechanism, particularly for the SERRS case, where the electronic structure of the adsorbate is reflected in the scattering. A SERRS spectrum that is identical to the solution resonance Raman spectrum of the molecule strongly supports an EM enhancement mechanism. Perturbation of the ground or excited states of the adsorbate by the metal surface would produce changes in band frequencies and intensities, respectively. The SERRS excitation profile provides information about the enhancement mechanism. The profile reflects the electronic resonances involved in the enhancement process. For SERS scattering, the profile should track the absorption spectrum (for Ag films or colloidal suspensions) or the reflectance spectrum (for electrodes) of the SERS active substrate. In the case of SERRS, the profile should track the product of the molecular absorption and the surface absorption in the absence of dipolar interactions that may shift the molecular resonance (and thus the excitation profile) to lower energies. Evidence for a chemical enhancement mechanism includes a strong dependence of relative band intensities or frequencies upon potential, an excitation profile which departs significantly from the profile expected based upon the molecular and metal resonances, and significant enhancement on surfaces that should not produce EM enhancement (e.g., smooth Ag). These characteristics can allow assessment of the dominant enhancement mechanism.

Descriptions of the theoretical treatment of SERS/SERRS have been presented to provide the reader with background for understanding the SERS literature. Many examples of SERS exist in the literature and the reader is referred to Seki[6] for a comprehensive list of references for papers published prior to 1986. Bibliographies which include references to SERS/SERRS studies published between October 1984 and May 1988 also exist.[8,36]

III. APPLICATIONS OF SERRS TO BIOLOGICAL SYSTEMS

In this section, the literature regarding experimental aspects of SERRS will be reviewed. A brief review of dye systems provides examples of the applications of SERRS and of the types of information that can be obtained from SERRS. Examination of biological chromophores about which multiple investigations were reported will be emphasized. These will include proteins and larger biological assemblies (membranes).

From the theoretical review presented, it is apparent that the definition of SERRS becomes vague for some enhancement mechanisms. In this review, experiments involving adsorbed species that possess electronic transitions near the laser excitation wavelengths are regarded as SERRS experiments. No discrimination between SERRS and surface resonance Raman scattering (SRRS) is made. The description of SRRS is often made for surfaces that do not give rise to enhancement via surface plasmons (e.g., smooth Ag or other metal surfaces). Nevertheless, significant chemical enhancement may occur on these surfaces, and small enhancement factors (<50) are calculated for adsorbates on smooth metal surface.[27,28] These enhancements involve bulk plasmons.

A. Summary of Application of SERRS to the Study of Dyes

A number of different types of dyes have been studied by SERRS. These include azo dyes,[4,62–67] phthalocyanines,[22,68–70] porphyrins,[71–75] fluoresceins,[76,77] rhodamine 6G,[22,78–84] viologens,[85–88] and cyanine dyes.[89–93] Results from these studies provide an understanding of the combined SERS and RR effects and, therefore, the salient conclusions are presented here.

Based upon the dyes studies referenced above, a number of generalizations may be made regarding the experimental manifestations of SERRS. Enhancement factors of 10^3 are typically reported.[20,63–66,80] This is lower than the value of 10^6 which is common for SERS, in accordance with predictions made by Metiu[13] and Weitz et al.[20] This lower enhancement may reflect the diminished lifetime of the excited state involved in the molecular resonance process.[13,20,22] Higher enhancement factors (10^6 and 10^{10}), however, were observed for some systems.[22,82] Furthermore, the enhancement factor decreases as the excitation wavelength is tuned into the molecular resonance.[63] The involvement of the molecular resonance in limiting the enhancement is illustrated by the same wavelength dependence of relative intensities observed in both the RR and SERRS spectrum.[75] The molecular resonance is also reflected in the SERRS excitation profile.[4] These observations and the similarity of RR and SERRS spectra[63,85,87] support the involvement of EM enhancement in these systems.

Band shifts observed in the spectra provide information about the orientation and points of interaction of the adsorbate at the metal surface.[71,75,77] Significant quenching of fluorescence is observed[20,76–82,89,90] which allows the observation of resonance spectra that are otherwise unobtainable. Fluorescence enhancement is also observed for some adsorbates.[20,91] Finally, metal incorporation[71,72] and metal replacement[72–74] appear to occur for those molecules that can form metal complexes in solution.

B. Biological Systems

1. Heme Proteins

Cytochrome c.   The most successful demonstration of the success of analysis of proteins by SERRS has come from the application to the heme-containing proteins known as cytochromes. The most widely studied of these proteins is cytochrome c (cyt c). Cytochrome c is relatively small (MW $\approx$ 13,000 ca. 31 Å sphere[94]) and has been well characterized: complete amino acid sequences were determined for over 50 species. The extent to which this protein has been characterized makes it well suited for the study of effects of adsorption on the protein structure. Furthermore, the heme chromophore, which possesses strong absorption bands in the visible region, has been well-characterized by resonance Raman spectroscopy.[95] The high apparent symmetry of the heme (near D_{4h}) makes possible the assignment of observed modes to specific symmetries. Excitation into the intense Soret band (near 400 nm) produces selective enhancement of totally symmetric (A_{1g}) modes through an "A-term" enhancement mechanism.[56] Excitation into the Q-bands (500–550 nm) produces enhancement of in-plane a_{2g}, b_{1g} and

b_{2g} modes through vibronic coupling of the Q-bands to the higher energy Soret absorption band (B-term enhancement.[56])

The Raman spectrum of heme chromophores is very sensitive to environmental influences. The redox state of the heme iron is reflected in the position of the band observed at ca. 1370 cm^{-1} in the oxidized heme and ca. 1360 cm^{-1} in the reduced heme. Additionally, the spin state of the Fe complex is very sensitive to the state of coordination and steric hindrance in the vicinity of the heme. Several bands are observed in the resonance Raman spectrum, and these bands provide information about the spin state (high or low) and coordination state (5 or 6) of the heme.

Cytochrome c possesses a histidine nitrogen and methionine sulfur as its two axial ligands and is thus 6 coordinate in its native state. The associated ligand field dictates that the iron complex exists in the low spin state. Perturbation of the position of the ligands or cleavage of the heme from the protein results in a change in the spin state from low to high. An indication of the spin state of the heme is given in the resonance Raman spectrum. Bands at 1502 cm^{-1}, 1584 cm^{-1} and 1635 cm^{-1} in low spin ferricyt c and at 1493, 1590 and 1620 cm^{-1} in ferrocyt c shift upon change in spin and coordination states. Thus, the positions of these bands in the resonance Raman spectrum provide valuable information about the immediate environment of the heme within the protein.

The redox behavior of cyt c is well established. The protein functions as a soluble electron transfer protein within the organism. The formal reduction potential of cyt c was reported as +0.260 V (vs NHE).[96] Reed and Hawkridge[97] reported the reduction of native cyt c at a Ag electrode by direct electron transfer. The authors report the effects of contaminants in the preparation (presumably deaminated and polymeric forms) upon the electrode kinetics. When the contaminants are present, the reduction potential shifts negative by approximately 150 mV, and the reduction becomes irreversible. The report of direct electron transfer of the native cyt c is evidence that (under certain conditions) the protein retains its native conformation upon interaction with Ag. Taking advantage of this knowledge, the authors monitored spectroelectrochemically the reduction of cyt c (by following the 1370 cm^{-1} SERRS band) and verified that the functional state of the protein was preserved by comparison of the observed reduction potential to that reported for the native protein in solution.

The wealth of available resonance Raman data and the high quality spectrum obtained from cyt c make it one of the first biological systems to be examined by SERRS. Cotton et al.[98] reported the SERRS spectrum of cyt c and myoglobin adsorbed on a Ag electrode (from 10^{-6} M solutions). Comparison of the solution (RR) and SERRS spectra with excitation at 514.5 nm shows great similarity, suggesting that the protein is not extensively denatured at the Ag surface.

More detailed studies using SERRS of the functional state of cyt c adsorbed at Ag surfaces were reported more recently.[99,100] Smulevich and Spiro[99] reported evidence that adsorption of heme proteins on Ag colloids causes denaturation. This study examined the SERRS spectra of cyt c, hemoglobin and cyt b (at 10^{-6} M concentrations) on Ag colloids by using 413.1 nm excitation. By comparison of the RR spectra of the protein in H_2O, the RR spectrum of Fe^{3+}, protoporphyrin IX μ-OXO dimer (in CH_2Cl_2) and the protein SERRS spectrum, the authors concluded that for each of the proteins studied, the SERRS active species represented, in part, the μ-oxo dimer formed from heme that was extracted from the protein. Spin state changes (from low to high) are evident in the protein SERRS spectra and are consistent with the formation of the μ-oxo dimer, which is high spin in nature. The release of heme from the cyt c requires cleavage of the covalent (thioether) attachments of the heme macrocycle to the surrounding protein matrix. This release was postulated to be mediated by Ag^+, which is known to cleave the heme from cyt c.[101] The Ag^+ may exist as a normal component of the colloidal preparation, associated with the metal surface.[102] The interaction of the heme propionate groups was suggested as promoting the extraction of the heme. Examination of the low frequency (150–450 cm^{-1}) SERRS bands of cyt c showed further evidence of this extraction, but only for the ferricyt c. This region, which is also very sensitive to the heme environment, showed no significant difference for ferrocyt c from that in the RR spectrum. On the basis of the evidence, the reduced cyt c may be more stable to the perturbing influence of the surface.

In a subsequent report, Hildebrandt et al.[100] provided convincing evidence that the spin state changes observed in the SERRS spectrum of cyt c (10^{-7}–10^{-8} M) on Ag colloids is not attributable to μ-oxo dimer formation (and hence, denaturation) but is a consequence of a reversible interaction of the protein with the Ag. This evidence comes from the observation of reversible change in the relative populations of high and low spin hemes with change in temperature. On Ag colloids at room temperature, a mixture of high and low spin heme was observed. Upon cooling of the colloid preparation to –100 °C, the population of high spin heme diminishes. At –196 °C, only low spin heme was observed. This trend was reversed upon warming back to room temperature. The nature of this spin state equilibrium is attributable to the perturbing influence of the Ag/electrolyte interface upon the ligand field strength in the heme complex. For O_h symmetry around the heme, the Fe d_{z^2} and $d_{x^2-y^2}$ orbitals are split in the ligand field to higher energy than the remaining d orbitals. The difference between the energy gap and the electron pairing energy determines the spin state of the complex. In solution, the energy gap is sufficiently large that only the d_{xy}, d_{xz} and d_{yz} orbitals are populated. Upon interaction with the surface, the energy gap is diminished so that the higher energy d orbitals become thermally populated. The influence of the Ag/electrolyte interface is postulated to be mediated through electrostatic interactions.

SERRS enhancement factors were estimated at a variety of wavelengths, and range from 8×10^4 (at 406.7 nm) to 1.1×10^6 (at 514.5 nm). The SERRS spectra also provide evidence for the retention of the same selection rules dictated by A-term and B-term enhancement as observed for solution scattering. This supports the hypothesis that no significant perturbation of the electronic structure of the heme occurs. The mechanism of enhancement is purely electromagnetic.

More insight into the nature of the cyt c SERRS active species was gained by examining the potential dependent behavior of the protein at the Ag electrode.[103] When the cyt c was adsorbed onto the electrode at an applied potential of −0.4 V (vs SCE), the SERRS spectra at potentials between −0.4 and +0.1 V are identical to the solution RR spectra of ferri- and ferrocyt c (low spin only). This is in contrast to the case of adsorption at an applied potential of +0.1 V where a SERRS spectrum identical to that reported on the Ag colloid was observed (mixed high and low spin). This behavior is attributed to two different adsorbed states (States I and II) of the protein. In State I, the protein is weakly adsorbed to the Ag surface, and in a fashion that does not significantly perturb the ligand field splitting in the heme. Thus, State I gives rise to SERRS spectra that are indistinguishable from the RR spectra and a redox potential that is identical to that in solution. State II represents more strongly adsorbed protein, and the protein is perturbed to produce an increase in ligand field splitting which gives rise to the temperature dependent equilibrium between the low and high spin states that were reported to explain the behavior on the Ag colloids. The spectroelectrochemically determined reduction potential for State II is shifted to a more negative value ($\sim$ −0.35 V).

The physical difference between States I and II appears to be the orientation of the protein at the Ag surface. Irradiation of State I at + 0.1 V leads to conversion to State II with a time constant of ca. 2 hr, a time constant that was also characteristic for desorption of cyt c from the electrode surface. Furthermore, a difference in orientation is consistent with the difference in observed reduction potentials for States I and II. An orientation that permits favorable electron transfer is consistent with State I whereas one that inhibits direct electron transfer is consistent with State II.

As described above, the spin state changes are attributable to electrostatic perturbation of the ligand field strength by the field within the electrical double-layer at the Ag/electrolyte interface. This is supported by the observation that spin state equilibria are sensitive to the ionic strength of the electrolyte. In our laboratory, SERRS studies of cyt c and cyt P-450 as a function of laser irradiation time have demonstrated that photothermal degradation of the proteins occurs under certain conditions.[104] Submersion of the electrode into liquid nitrogen effectively eliminated photodegradation on a Ag electrode while maintaining the native state of cyt c. This procedure resulted in stable SERRS spectra during prolonged irradiation.

Hemoglobins and Myoglobins.   Several reports were made of the retention of the native state of hemoglobin and myoglobin adsorbed at Ag colloidal surfaces.[105–108] Figure 3 shows a direct comparison of the SERRS spectrum of myoglobin on a Ag electrode at 77 °K with its RR spectrum in solution at room temperature.[108] The SERRS spectrum is nearly identical to the RR spectrum. The solution RR spectrum is characteristic of 6 coordinate, high spin heme ($v_3 = 1482$ cm^{-1}, $v_2 = 1567$ cm^{-1}) in the fully oxidized state ($v_4 = 1373$ cm^{-1}). Similar band

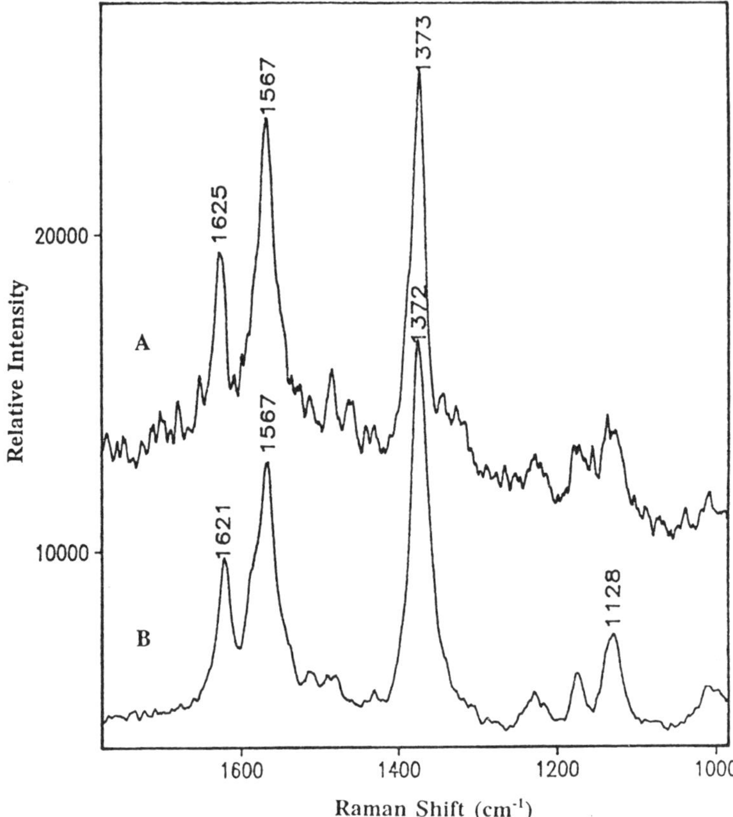

Figure 3. RR and SERRS spectra of myoglobin: (A) in solution (concentration = 85 mM) at room temperature and (B) on a Ag electrode at 77 K. The laser excitation wavelength was 413.1 nm and the power was 25 mW. The spectra were recorded on an intensified photodiode array detector and are a composite of 25 accumulated scans with an integration time of 1 scan/sec.).[108]

positions are noted in the SERRS spectrum, but the 1625 cm^{-1} band, attributed to the vinyl mode, is downshifted to 1621 cm^{-1}. This small shift may reflect a direct interaction of the vinyl group with the surface.

In another study demonstrating native structure of a heme protein on a Ag surface, de Groot et al.[105] demonstrated that the SERRS spectrum (excitation wavelength = 406.7 nm) of oxyhemoglobin (HbO$_2$, 10^{-6}–10^{-7} M) gives no indication of the presence of a high spin component, in contrast to the results of Smulevich and Spiro.[99] The presence of only low spin heme strongly supports the retention of native state at the Ag surface. The major difference between the SERRS experiments reported by Smulevich et al.[99] and de Groot et al.[105] is the procedure used for preparation of the Ag colloid. Preparation of the colloid by citrate reduction produces the low spin SERRS spectrum reported by de Groot et al. whereas borohydride reduction was used by Smulevich et al. Comparison of the SERRS spectrum of the oxyhemoglobin to that of the Fe^{3+} protoporphyrin IX confirms that no significant amount of μ-oxo dimer is present on the Ag surface. No SERRS enhancement factors were reported but the observation of strong scattering from the low concentration of hemoglobin suggests that a surface-enhancement mechanism is operative. The enhancement of the totally symmetric vibrational modes when Q-band excitation was used suggests that a surface interaction occurs which lowers the symmetry of the heme.

Taking advantage of the well characterized functional activity of hemoglobin, one group confirmed the retention of native state of hemoglobin (1×10^{-7} M concentration) on Ag colloids (citrate reduction).[106] Using carbonylhemoglobin (HbCO), they found that the sequential photolysis (using laser irradiation at 441.6 nm) to deoxyhemoglobin (Hb) was followed by binding of either CO or O_2 introduced into the SERRS cell. Additionally, HbCO was observed to form directly from HbO_2 by the introduction of CO into the cell. These changes are reflected in the SERRS spectrum by changes in the relative intensities of bands at 1370 (assigned to HbO2), 1375 (HbCO) and 1358 cm^{-1} (Hb). The SERRS spectrum reported for HbCO was identical to its RR spectrum. The low frequency region of Hb shows identical peaks at 215 cm^{-1} in the SERRS and RR spectra. This peak has been assigned to an Fe-NH mode, which is sensitive to the quaternary structure of the protein. These observations are strong evidence that the adsorbed hemoglobin retains its functional ability in the adsorbed state.

The difference between the colloidal preparations using citrate and borohydride reductants was also explored by de Groot et al.[106] A comparison of the SERRS spectra of hemoglobin obtained from both preparations confirms the hypothesis that the difference in the functional activity reported by de Groot et al.[106] and Smulevich et al.[99] is attributable to differences in the colloid preparation procedure. The authors conclude that Ag colloids prepared by using borohydride reduction apparently provide a harsher chemical environment and raised doubts about the applicability of borohydride for SERRS studies of biological systems.

Cytochrome c. Cytochrome c_3 is a low molecular weight electron transport protein (MW $\sim$ 13000) found in high concentration in sulfate-reducing bacteria. Despite its small size, the protein contains 4 heme groups, each having a different macroscopic formal reduction potential. A SERRS investigation of the redox properties of cyt c_3 obtained from several different sources and adsorbed on a Ag electrode (no bulk cytochrome present) was reported by Niki et al.[108] Using differential pulse voltammetry, the authors showed that the formal reduction potential of the cytochrome is an average of the very similar macroscopic formal potentials of each of the individual hemes. By monitoring the oxidation state marker in the SERRS spectrum of cyt c_3 (λ_{ex} = 514.5 nm), the formal reduction potentials of the cytochromes from the different sources were all determined to be at least 0.1 V more positive than those measured by using differential pulse polarography. The discrepancy was accounted for by consideration of the orientation of adsorbed cyt on the Ag surface. If the cyt adsorbs in a preferred orientation, only one of the four hemes is likely to produce a strong SERRS spectrum. The spectroelectrochemical experiment is therefore only monitoring the oxidation state of this particular heme. The spectroelectrochemically determined reduction potential is that of the single heme and not the formal reduction potential (determined by contribution from all four hemes) that is measured in the differential pulse experiment. Comparison of the spectroelectrochemically determined reduction potentials to the macroscopic formal potentials suggests that the heme that is interacting with the Ag surface is the one that possesses the most positive reduction potential (Heme 1). Yet, the spectroelectrochemically determined reduction potentials do not completely agree with the macroscopic formal potential of Heme 1, but are believed to correspond to the intrinsic reduction potentials of Heme 1 that would result in the absence of its interactions with the other hemes.

Cytochrome cd₁. Cytochrome cd_1 is the penultimate component in the electron transfer chain of denitrifying bacteria. It accepts electrons from cytochrome c-type donors and transfers them to nitrite, thereby acting as a nitrite reductase. The protein consists of two subunits of ca. 60,000 MW, each subunit possessing one c-type and one d_1-type heme.

Cotton et al.[109] investigated the application of SERRS to the study of reduced cytochrome cd_1 (from 3×10^{-6} M solution) on Ag electrodes. The different absorption spectra of the c-type and d_1-type hemes allow selective excitation by using 457.9 (c-type) and 514.5 nm (d_1-type). This selective excitation is demonstrated in the solution RR spectra of the cytochrome. Comparison of the SERRS and RR spectra shows significant differences, particularly in relative band intensities. Marker bands indicate that heme c is present as the ferrous low-spin complex. Differences in structures of hemes c and d_1 are reflected in the SERRS spectra. Differences in the SERRS spectra of cytochrome cd_1 isolated from different sources (*Paracoccus,* and *Pseudomonas*) agree with different physicochemical properties reported for cytochrome cd_1 from these two sources, although the precise nature of these differences cannot be inferred from the spectra.

Cytochromes P-450. The examples of application of SERRS to the study of cytochromes described thus far have relied upon either retention of the spin state character or the reduction potentials of the proteins in the adsorbed state to confirm the preservation of the functional activity. Another class of cytochromes designated as P-450 acts as enzymes in many organisms, catalyzing the oxidation of potentially toxic compounds[110] by using molecular oxygen. Resonance Raman spectroscopy was used to explore the effects of substrate binding upon the spectrum.[111] Changes in the RR spectrum are manifested primarily by transitions between low and high spin states.

Kelly et al.[112] examined the SERRS spectra of a number of cytochromes P-450 from different sources. These cytochromes possess different relative populations of low and high spin character in their native states. A comparison of the SERRS spectra of these proteins (at ca. 10^{-7} M concentrations) on Ag colloids (citrate reduced) by using both Soret (457.9 nm) and Q-band (514.5 nm) excitation allows the precise identification of the marker bands (spin state and oxidation state) in the SERRS spectra. Small differences (<5 cm^{-1}) are observed in the positions of these bands from P-450 from different sources. Differences between the SERRS spectra of the P-450 and P-420, the biological inactive form of the cytochrome, are also apparent. The SERRS spectrum of P-420 shows a band at 1627 cm^{-1} and a decrease in the relative intensity of bands around 1400 cm^{-1}.

In a subsequent report by the same research group,[113] SERRS on Ag colloids was used to investigate the biochemical differences between cytochromes P-450 PB$_{3a}$ and PB$_{3b}$, two cytochromes that share 97% sequence homology yet show different enzymic specificities. Differences in spin state character were attributed to differences in amino acid composition near the heme chromophore. Evidence is shown for spin state change in PB$_{3b}$ upon binding of the substrates benzphetamine and cyclohexane. These changes are more pronounced than those observed for PB$_{3a}$ upon substrate addition. No evidence for formation of P-420 was observed and the similarity of the SERRS spectra to the reported RR spectrum[111] and change in spin state upon substrate binding confirm the retention of native functional activity in the adsorbed state.

Retention of the functional integrity of cyt P-450 was more clearly demonstrated by Hildebrandt et al.[114] Using 406.7 nm excitation, these researchers observed changes in the RR spectrum from a low spin to mixed low/high spin state upon addition of substrate (benzphetamine). These changes were reproduced in the SERRS spectra of the protein (10^{-7}–10^{-8} M concentrations) on Ag colloids with and without added substrate. Evidence was also presented for a photoinduced change in P-450 to P-420 during laser irradiation. In cyt P-450, the heme exists (in the unbound state) as a 5-coordinate complex with a thiolate group as the axial ligand. This ligand causes the position of the oxidation state marker band to appear at lower frequencies than in cyt c in the reduced state (1343 cm^{-1} as compared to 1360 cm^{-1}). The observation of three bands around 1370 cm^{-1} in the SERRS spectrum of reduced P-450 (at 1343,1359 and 1370 cm^{-1}) suggests that native cyt P-450 is present together with cyt P-420. A laser irradiation dependent decrease in the 1343 cm^{-1} band in both the RR and SERRS experiments is evidence that this photolability is intrinsic to the protein and not a consequence of its adsorption on the Ag surface. The presence of the 1343 cm^{-1} band in the reduced cyt P-450 spectrum suggests that the thiolate axial ligand is not perturbed in the adsorbed state.

The native structure of Cyt P-450 was also preserved on a phospholipid-coated

Ag electrode at room temperature.[115] Although the SERRS bandshifts correlated with the RR spectra, the relative intensities of specific frequencies changed in the presence of adsorbed phospholipid. The strong interaction of Cyt P-450 with the coated lipid may result in a specific orientation of the protein at the electrode surface. This, in turn, leads to an enhancement of specific modes due to the surface selection rules.

2. Dye-Molecule Complexes

Avidin–HABA Complex. Several SERRS investigations have taken advantage of the strong SERRS scattering from dyes to investigate the interaction of these dyes with molecules of biological interest.

Ni and Cotton[116] have reported the SERRS spectrum of 2-(4'-hydroxyphenylazo)benzoic acid (HABA) at Ag electrodes. HABA binds very strongly to the protein avidin ($K_D = 5.8 \times 10^{-6}$ M), a tetrameric protein having a molecular weight of ca. 68,000 and which is known to bind the compound biotin with one of the highest affinities observed in nature ($K_D \sim 10^{-15}$ M). Changes in the absorption spectrum of HABA occur upon binding to avidin, and the changes allow the selective excitation of only the bound chromophore in the Raman experiment. Resonance Raman studies of the binding of HABA to avidin[117] and to bovine serum albumin (BSA)[117,118] were reported and provide a basis for the comparison of the SERRS results. Of the tautomeric forms available to HABA, the hydrazone predominates in avidin[117] whereas both the hydrazone and azo forms exist in BSA.[117,118]

The RR and SERRS spectra of the HABA-avidin complex show only minor differences. The presence of a weak band in both spectra is indicative of a small amount of azo form of the HABA bound in the protein. In the SERRS spectrum, this band is broadened and shifted to 1406 cm^{-1}, indicating that the protein containing HABA bound in the azo form may be more susceptible to perturbation by the metal surface. The SERRS spectrum of the complex differs from that of free HABA and is therefore attributable to protein bound HABA that is prevented from directly interacting with the Ag surface. Avidin, which is adsorbed to the Ag electrode in the absence of HABA, was demonstrated to bind HABA when added to the system. The resulting SERRS spectrum is identical to that of the complex formed by free avidin and HABA. The similarities of the RR and SERRS spectra of the avidin-HABA complex and the demonstration of binding of HABA to avidin that was previously adsorbed on Ag are strong evidence for the retention of the native configuration of avidin following adsorption to Ag.

Anthracycline-DNA Complexes. Anthracycline antibiotics are used in the treatment of cancer, and their biological activity is attributed to their formation of intercalation complexes with DNA. In these complexes, the chromophore is inserted into the stacked base pairs in the DNA. Using SERRS on Ag colloids,

Smulevich et al.[119,120] investigated the interactions of adriamycin (ADM), 11-deoxycarminomycin (DCM) and related model compounds (at 10^{-6} M concentrations) with DNA. Excitation at 457.9 nm produced strong spectra of the anthracycline compounds despite the strong fluorescence exhibited by these compounds in solution. The resulting SERRS spectra are identical to the previously reported RR spectra. Downshifts in SERRS bands at 490, 1207 and 1303 cm^{-1} the anthracycline-DNA complex spectrum indicate that the chromophore interacts with the Ag in an edge-on fashion. These same bands in the DNA complexes support a similar interaction. Furthermore, interaction with the Ag surface does not significantly perturb the complex. For the more symmetric model compounds, some totally symmetric modes decreased in intensity, whereas no change in B_1 modes occurred. These changes are apparently a consequence of the binding interaction. Intercalation with the DNA perturbs the excited electronic state involved in the resonance enhancement process. Those modes that are most affected are localized on a portion of the anthracycline that interacts directly with the DNA. Thus, the changes in relative intensities are consistent with the generally accepted model of intercalation of these compounds into DNA.

3. Membrane-Bound Systems

Purple Membranes.   The purple membrane (PM) of purple sulfur bacteria serves to channel incident light energy into the generation of a proton gradient across the membrane.[121] These membranes contain the chromophoric protein bacteriorhodopsin. The chromophore, retinal, is attached to the protein via a Schiff base and proton transport across the membrane is associated with isomerization of the retinal in a manner that is still the subject of active investigation. Of particular importance are the location of the retinal within the protein and the conformational changes that occur upon activation by light.[122,123]

Nabiev et al.[122,123] applied SERRS at Ag electrodes and colloids to study the organization of the purple membrane from *Halobacterium halobium*. Figure 4 shows the SERS (A,B) and RR (C) spectra of the PM from *Halobacterium halobium*. SERS spectra of PM in Figure 4-I were recorded from unaggregated Ag sol, whereas Figure 4-II were from aggregated Ag sol. The former spectrum results from short-range, chemical enhancement. The chemical interaction of the protein with the surface results in denaturation and no photocycle intermediate can be seen. On the other hand, the spectra recorded for an aggregated sol result from long range, EM enhancement. The spectra are similar to the RR spectra of the native protein and the photocycle intermediate is formed on the surface (see Fig. 4, II A) as evidenced by an increase in the band at 1570 cm^{-1}. The latter supports the presence of native protein on the surface. The SERRS spectra of purple membranes and *Halobacterium* cell envelopes on Ag electrodes are dominated by interference attributed to graphitic carbon that forms during laser irradiation from adsorbed CO_2.[120] Other sharp bands in the spectra were attributed to aromatic and carboxyl

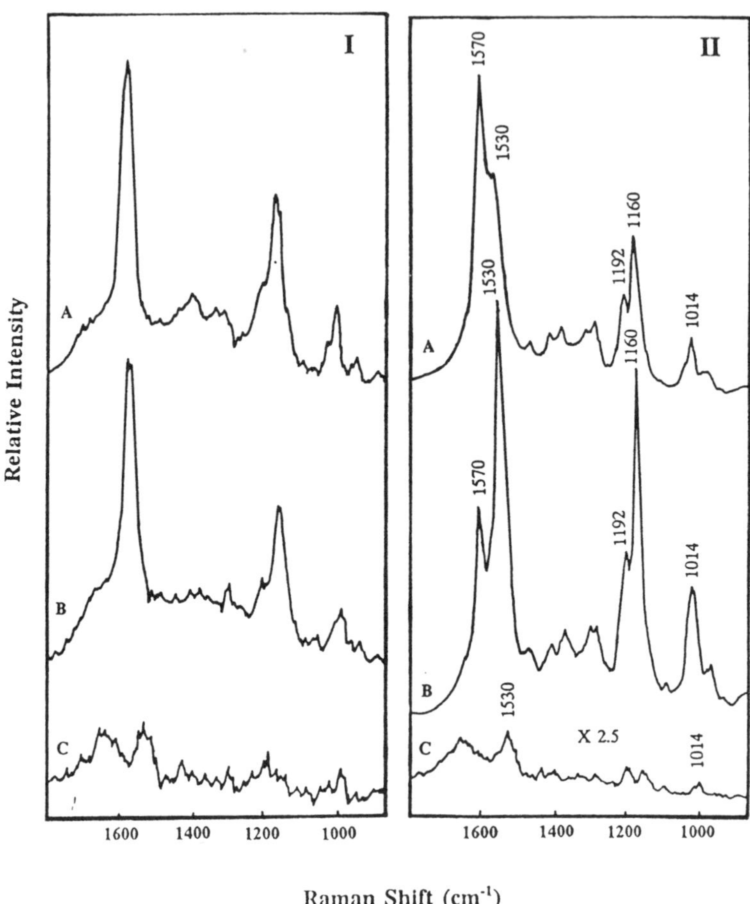

Raman Shift (cm⁻¹)

Figure 4. SERS (A,B) and RR (C) spectra of the PM from *Halobacterium halobium* on unaggregated (I) and aggregated (II) Ag hydrosols. λ_{ex}: I-A = 476.5 nm; I-B and C = 541.5 nm; II-A = 457.9 nm; II-B and C = 514.5 nm.[123]

modes of the amino acids. Weak bands were assigned to the retinal chromophore by comparison to the RR spectrum of the purple membrane. Better SERRS spectra were obtained on Ag colloids (excitation wavelength not stated).

Lutz et al.[125] described the changes that occur in the RR spectrum of the similar polyene compound β-carotene. The most significant change upon isomerization of the all-*trans* form to the 15,15′-*cis* form is an upshift in the C=C band at 1530 cm⁻¹ (in the all-*trans* form) to 1540 cm⁻¹. Other changes include small band shifts and changes in relative intensities. Despite the statement to the contrary, the differences between the RR and SERRS spectra (on Ag colloids) of the purple membranes described above do not convincingly rule out the possibility of isomerization of the retinal.

By comparison of the SERRS intensities from chromoproteins that were reconstituted with retinal analogs of different polyene lengths, an estimate of the distance between the external surface of the purple membrane and the retinal Schiff base linkage was made. The authors assumed that a chemical mechanism is operative in producing the SERRS of retinal at the Ag surface. They estimated the distance of the chromophoric Schiff base attachment from the surface as 6–9 Å.

Bacterial Photosynthetic Membranes.   Picorel et al. used SERRS to probe the distribution of carotenoids in the photosynthetic membranes isolated from *Rhodospirillum rubrum*, a photosynthetic bacterium as shown in Figure 5.[128–130] Chromatophores (membrane preparations having cytoplasmic side-out orientations) and spheroplasts (membranes having periplasmic side-out orientations) were isolated from the F-24 strain (a reaction centerless mutant). RR spectra of these preparations obtained with excitation wavelengths between 457.9 and 568.2 nm are intense, arising exclusively from the carotenoid spirilloxanthin. At much lower concentrations, SERRS signals arising from spirilloxanthin were observed for the chromatophore preparation when present during the electrochemical roughening of the Ag electrode as well as when added after the roughening procedure. The three major peaks in the RR spectrum were observed in the SERRS spectrum at identical frequencies. The enhancement mechanism appears to be controlled by the molecular resonance because the same trends in relative band intensities (as a function of excitation wavelength) were observed in the RR and SERRS spectra. No significant SERRS enhancement of the spirilloxanthin bands was observed in the spheroplast preparation. This suggests that the spirilloxanthin is distributed asymmetrically with respect to the membrane and is predominantly on the cytoplasmic side. Other explanations for the absence of SERRS enhancement in the spheroplast were considered: the presence of cell wall around the spheroplast membrane, low chromophore density in the spheroplast membrane, lack of spheroplast adsorption onto the Ag electrode and differences in surface charge affecting the adsorption of the spheroplasts on the Ag. These possibilities were ruled out by various experiments. The asymmetric distribution of the spirilloxanthin as probed by SERRS supports previous predictions that have been made by using antibody[131] and proteinase[131,132] treatments.

Photosynthetic Reaction Centers.   Cotton and Van Duyne[126] applied SERRS at Ag electrodes to the study of the photosynthetic reaction center from *Rhodopseudomonas sphaeroides* R-26 carotenoidless mutant. The reaction center serves as the apparatus through which light energy is converted to chemical energy in the photosynthetic organism. The reaction center of *R. sphaeroides* has been well characterized and is a large assembly of 3–4 proteins containing four bacteriochlorophylls (BChl), two bacteriopheophytins (BPh) and a quinone.[127]

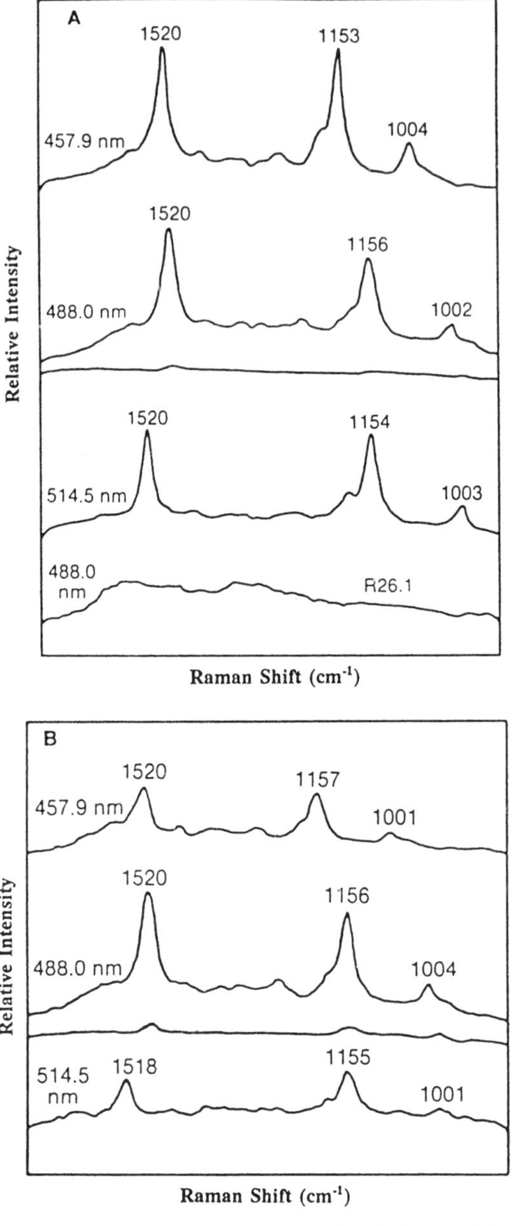

Figure 5. SERRS spectra of chromatophores isolated from wild-type and R 26.1 mutant cells of *Rhodobacter sphaeroides* excited at the 457.9, 488.0, and 514.5 nm: (A) Light grown cells; (B) dark-grown cells. The trace accompanying the spectra excited at 488.0 nm is a RR spectrum of the sample adsorbed on the Ag electrode prior to the oxidation-reduction cycle. The spectrum at the bottom of (A) is that of chromatophores from R 26.1. The carotenoid concentration was 0.2 (A) and 0.3 (B) μg/mL. The spectra of (B) are magnified 2-fold. The buffer was 20 mM HEPES (pH 7.5)-100 mM Na_2SO_4.[130]

At a solution concentration of 1×10^{-6} M and 457.9 nm excitation, bands were observed in the spectrum which were attributed to bacteriochlorophyll (BChl). The spectrum is similar to the RR spectrum of BChl/pyridine (1:1) in CH_2Cl_2, with some minor differences. Furthermore, the spectra are potential dependent. Spectra taken at -0.7 V versus SCE are more intense than those at 0 V. When the excitation wavelength was changed to 530.8 nm, a spectrum identical to BPh was observed. In contrast to the BChl spectrum, this spectrum is most intense at 0 V and decreases at more negative potentials. The different potential dependencies may be due to potential dependent changes in orientation or conformation of the protein complex on the surface.

IV. SUMMARY OF SERRS APPLICATION TO BIOLOGICAL SYSTEMS

The review of applications of SERRS to the study of biological systems emphasizes the power of this technique. Most significant is the demonstration that many protein systems appear to retain their functional activity when adsorbed on the Ag surface. This was shown to be the case for all of the heme proteins. These proteins are generally small or possess a high chromophore density which allows a high density of chromophores to be positioned close to the Ag surface and thus a strong SERRS spectrum is produced. Concentrations that were reported were always less than ca. 10^{-6} M[98,99,105,106,108,112,119,120,126] and as low as 10^{-8} M.[100,114] Usually small volumes of material are required for the analysis (e.g., 10 μL).[112] These requirements make practical the analysis of precious biological material. Furthermore, the quenching of fluorescence allows the observation of high quality Raman spectra.[119,120]

Both Ag colloids and electrodes were used successfully in these applications. The method of colloid preparation appears to be important. Preparations that used sodium borohydride as the reducing agent give SERRS spectra that are characteristic of the perturbed or denatured systems[99,106] whereas colloids prepared with citrate produce spectra of the native systems. The pH of the colloid preparation also appears to be critical.[112,113] The use of the Ag electrode provides an additional advantage in allowing potential control. For some systems, the ability to control the potential is a prerequisite for the demonstration of retention of the functional state of the proteins on the Ag surface.[100,107]

REFERENCES

1. Morris, M. D.; Wallan, D. J. *Anal. Chem.* **1979**, *51*, 182A.
2. Gustavson, T. L.; Lytle, L. F. *Anal. Chem.* **1982**, *54*, 634.
3. Demas, J. N.; Keller, R. A. *Anal. Chem.* **1985**, *57*, 538.
4. An excellent description of the events associated with the first observation of SERS is provided

by Richard Van Duyne In *Chemical and Biochemical Applications of Lasers*, Moore, C. B. Ed.; Academic: New York, 1978; Vol. 4, p. 101.

5. A) Jeanmaire, D. L.; Van Duyne, R. P. *J. Electroanal. Chem.* **1977**, *84*, 1. B) Albrecht, M. G.; Creighton, J. A. *J. Am. Chem. Soc.* **1977**, *99*, 5215.

6. Seki, H. *J. Elec. Spectrosc. Rel. Phenom.* **1986**, *39*, 289.

7. Efrima, S. In *Modern Aspects of Electrochemistry*, Conway, B. E.; White, R. E.; Borkris, J. O'M. Eds.; Butterworths: New York, 1985; No. 16, p. 253.

8. A) Chang, R. K. In *CRC Critical Reviews in Solid State and Material Sciences*, **1984**, *12*(1), 1. B) Chang, R. K. *Ber. Bunsenges. Phys. Chem.* **1987**, *91*, 296.

9. Howard, M. W.; Cooney, R. P.; McQuillan, A. J. *J. Raman Spect.* **1980**, *9*, 273.

10. Mahoney, M. R.; Howard, M. W.; Cooney, R. P. *Chem. Phys. Lett.* **1980**, *71*, 59.

11. Cooney, R. P.; Mahoney, M. R.; Howard, M. W. *Chem. Phys. Lett.* **1980**, *76*, 448.

12. Wokaun, A. *Solid State Phys.* **1984**, *38*, 223.

13. Metiu, H. *Prog. Surf. Sci.* **1984**, *17*, 153.

14. Efrima, S.; Metiu, H. *J. Chem. Phys.* **1979**, *70*, 1939.

15. Otto, A. In *Light Scattering in Solids IV*, Cardona, M.; Güntherodt, G. Eds.; Springer-Verlag: NY, 1984, p. 289.

16. Wang, D. S.; Kerker, M. *Phys. Rev. B* **1982**, *25*, 2433.

17. Chew, H.; Wang, D. S. *Phys. Rev. Lett.* **1982**, *49*, 490.

18. Kotler, Z.; Nitzan, A. *J. Phys. Chem.* **1982**, *86*, 2011.

19. Gersten, J. I.; Nitzan, A. *J. Chem. Phys.* **1981**, *75*, 1139.

20. Weitz, D. A.; Garoff, S.; Gersten, J. I.; Nitzan, A. *J. Phys. Chem.* **1983**, *78*, 5324.

21. Efrima, S. *J. Phys. Chem.* **1985**, *89*, 2843.

22. Zeman, E. J.; Carron, K. T.; Schatz, G. C.; Van Duyne, R. P. *J. Chem. Phys.* **1987**, *87*, 4189.

23. Appropriate reviews suggested by Metiu[10] are a) Behringer, J. In *Molecular Spectroscopy*; The Chemical Society: London, 1974, Vol. 2.; b) Behringer, J. In *Raman Spectroscopy*; Szymanski, H. A. Ed.; Plenum Press: NY, 1967; c) Spiro T. G.; Stein, P. *Ann. Rev. Phys.* **1977**, *28*, 501.

24. Chance, R. R.; Prock, A.; Silbey, R. *J. Chem. Phys.* **1974**, *60*, 2744.

25. Chance, R. R.; Prock, A.; Silbey, R. *J. Chem. Phys.* **1974**, *60*, 2184.

26. Chance, R. R.; Prock, P.; Silbey, R. *Phys. Rev. A* **1975**, *12*, 1448.

27. Efrima, S.; Metiu, J. *J. Chem. Phys.* **1979**, *70*, 1602.

28. Efrima, S.; Metiu, J. *J. Chem. Phys.* **1979**, *70*, 2297.

29. Feynmann, R. P.; Leighton, R. B.; Sands, M. *The Feynmann Lectures on Physics*; Addison-Wesley: Reading, MA, 1963; Vol. 1, p. 33.

30. Ferrell, T. L.; Callcott, T. A.; Warmack, R. J. *Am. Scientist* **1985**, *73*, 344.

31. Raether, H. *Excitation of Plasmons and Interband Transitions by Electrons*; Springer-Verlag; Berlin and New York, 1980.

32. Kerker, M.; Wang, D. S.; Chew, H. *Appl. Opt.* **1980**, *19*, 3373.

33. Gersten, J. I.; Nitzan, A. *J. Chem. Phys.* **1980**, *73*, 3023.

34. Murray, C. In *Surface-Enhanced Raman Scattering*; Chang, R. K.; Furtak, T. E., Eds.; Plenum Press: New York, 1982; p. 203.

35. Moskovits, M. *J. Chem. Phys.* **1982**, *77*, 4408.

36. Creighton, J. A. In *Advances in Spectroscopy*, Clark, R. J. H.; Hester, R. E. Eds.; John Wiley and Sons: New York, 1988, p. 37.

37. Moskovits, M.; Suh, J. S. *J. Phys. Chem.* **1984**, *88*, 5526.

38. Creighton, J. A. *Surf. Sci.* **1983**, *124*, 209.

39. Creighton, J. A. *Surf. Sci.* **1985**, *158*, 211.

40. Murray, C. A.; Allara, D. L.; Rhinewine, M. *Phys. Rev. Lett.* **1981**, *46*, 57.

41. Cotton, T. M.; Uphaus, R. A.; Möbius, D. *J. Phys. Chem.* **1986**, *90*, 6071.

42. Pettinger, B.; Wenning, U.; Kolb, D. M. *Ber. Bunsenges. Phys. Chem.* **1978**, *82*, 1136.

43. Aussenegg, F. R.; Lippitsch, M. E. *Chem. Phys. Lett.* **1978**, *59*, 214.

44. Abe, H.; Manzel, K.; Schulze, W.; Moskovits, M.; DiLella, D. P. *J. Chem. Phys.* **1981**, *74*, 792.

45. Gersten, J. I.; Birke, R. L.; Lombardi, J. R. *Phys. Rev. Lett.* **1979**, *43*, 147.

46. Persson, B. N. J. *Chem. Phys. Lett.* **1981**, *82*, 561.

47. Otto, A.; Billman, J.; Eickmans, J.; Erturk, U.; Pettenkofer, C. *Surf. Sci.* **1984**, *138*, 319.

48. Otto, A.; Pockrand, I.; Billman, J.; Pettenkofer, C. In *Surface-Enhanced Raman Scattering*; Chang, R. K.; Furtak, T. E. Eds.; Plenum Press: New York, 1982, p. 147.

49. Watenabe, T.; Kawanabe, O.; Honda, K.; Pettinger, B. *Chem. Phys. Lett.* **1983**, *102*, 565.

50. Furtak, T. E.; Roy, D. In *Advances in Laser Science II;* Lapp, M.; St Walley, W.C.; Kenney-Wallace, G.A. Eds.; American Institute of Physics Conf. Proceedings, NY 1987; No. 160, 452.

51. Roy, D.; Furtak, T. E. *Phys. Rev.* **1986**, *34*, 5111.

52. Corn, R. M.; Philpott, M. R. *J. Chem. Phys.* **1984**, *81*, 4138.

53. Adrian, F. J. *J. Chem. Phys.* **1982**, *77*, 5302.

54. Lippitsch, M. E. *Phys. Rev. B.* **1984**, *29*, 3101.

55. Lombardi, J. R.; Birke, R. L.; Lu, T. H.; Xu, J. *J. Chem. Phys.* **1986**, *84*, 4174.

56. Tang, J.; Albrecht, A. C. In *Raman Spectroscopy*; Szymanski, H.A. Ed.; Plenum Press: New York 1970; Vol. 2, p. 33.

57. Lu, T. H. Ph.D. Dissertation, City University of New York, 1987.

58. Fritz, H. P. *Adv. Organomet. Chem.* **1984**, *1*, 239.

59. Gao, P.; Weaver, M. J. *J. Phys. Chem.* **1985**, *89*, 5040.

60. Paterson, M. L.; Weaver, M. J. *J. Phys. Chem.* **1985**, *89*, 5046.

61. Moskovits, M.; DeLella, D. P. *J. Chem. Phys.* **1980**, *73*, 6068.

62. Campbell, J. R.; Creighton, J. A. *J. Electroanal. Chem.* **1983**, *143*, 353.

63. Bachackashvilli, A.; Katz, B.; Priel, Z.; Efrima, S. *J. Phys. Chem.* **1984**, *88*, 6185.

64. Siiman, O.; Lepp, A.; Kerker, M. *J. Phys. Chem.* **1983**, *87*, 5319.

65. Siiman, O.; Lepp, A.; Kerker, M. *Chem. Phys. Lett.* **1983**, *100*, 163.

66. Siiman, O.; Smith, R.; Blatchford, C.; Kerker, M. *Langmuir* **1985**, *1*, 90.

67. Ni, F.; Cotton, T. M. *J. Raman Spectrosc.* **1988**, *19*, 429.

68. Aroca, R.; Loufty, R. O. *J. Raman Spectrosc.* **1982**, *12*, 262.

69. Kötz, R.; Yeager, E. *J. Electroanal. Chem.* **1980**, *113*, 113.

70. Loutfy, R. O. *Can. J. Chem.* **1981**, *59*, 549.

71. Cotton, T. M.; Schultz, S. G.; Van Duyne, R. P. *J. Am. Chem. Soc.* **1982**, *104*, 6528.

72. Shoji, K.; Kobayashi, Y.; Itoh, K. *Chem. Phys. Lett.* **1983**, *102*, 179.

73. Kobayashi, Y.; Itoh, K. *J. Phys. Chem.* **1985**, *89*, 5174.

74. Takenaka, A.; Takauchi, S.; Kobayashi, Y.; Itoh, K. *Surf. Sci.* **1985**, *158*, 359.

75. Sanchez, L. A.; Spiro, T. G. *J. Phys. Chem.* **1985**, *89*, 763.

76. Chen, C. Y.; Davoli, I.; Ritchie, G.; Burstein, E. *Surf. Sci.* **1980**, *101*, 363.

77. Hildebrandt, P.; Stockburger, M. *J. Raman Spectrosc.* **1986**, *17*, 55.

78. Weitz, D. A.; Garoff, S.; Gramila, T. J. *Optics Let.* **1982**, *7*, 168.

79. Garoff, S.; Stevens, R. B.; Hanson, C. D.; Sorenson, G. K. *Optics Comm.* **1982**, *41*, 257.

80. Weitz, D. A.; Garoff, S.; Gersten, J. I.; Nitzan, A. *J. Electron. Spectrosc. Rel. Phen.* **1983**, *29*, 363.

81. Hildebrandt, P.; Stockburger, M. *J. Phys. Chem.* **1984**, *88*, 5935.

82. Pettinger, B.; Gerolymatou, A. *Surf. Sci.* **1985**, *156*, 859.

83. Chance, R. R.; Prock, A.; Silbey, R. *Advan. Chem. Phys.* **1978**, *37*, 1.

84. Pettinger, B. *Chem. Phys. Lett.* **1984**, *110*, 576.

85. Ohsawa, M.; Nishijima, K.; Suëtaka, W. *Surf. Sci.* **1981**, *104*, 270.

86. Lu, T.; Cotton, T. M. *J. Phys. Chem.* **1987**, *91*, 5978.

87. Feng, Q.; Cotton, T. M. *J. Phys. Chem.* **1986**, *90*, 983.

88. Lu, T.; Birke, R. L.; Lombardi, J. R. *Langmuir* **1986**, *2*, 305.

89. Kneipp, K.; Hinzmann, G.; Fassler, D. *Chem. Phys. Lett.* **1983**, *99*, 503.

90. Kneipp, K.; Fassler, D. *Chem. Phys. Lett.* **1984**, *106*, 498.

91. Gao, X.; Wan, C.; He, T.; Li, J.; Xin, H.; Liu, F. *Chem. Phys. Lett.* **1984**, *112*, 465.

92. Kim, J.-H.; Cotton, T. M.; Uphaus, R. A.; Möbius, D. *J. Phys. Chem.* **1989**, *93*, 3713.

93. Schlegel, V., M.S. Thesis, Univ. of Nebraska at Lincoln, 1986.

94. Dickerson, R. E.; Kopka, M. L.; Weinsierl, J. E.; Varnum, J. C.; Eisenberg, D.; Margoliash, E. In *Structure and Function of Cytochromes*, Okunuki, K.; Kamen, M.D.; Sekuzu, T. Eds.; Univ. of Tokyo Press: Tokyo, 1986; 225.

95. Spiro, T. G. *Acc. Chem. Res.* **1974**, *1*, 339.

96. Cusanovich, M. A. In *Bioorganic Chemistry*, Van Temelen, E. E. Ed.; Academic Press: New York, 1978; Vol. 4, p. 117.

97. Reed, D. E.; Hawkridge, F. M. *Anal. Chem.* **1987**, *59*, 2334.

98. Cotton, T. M.; Schulktz, S. G.; Van Duyne, R. P. *J. Am. Chem. Soc.* **1980**, *102*, 7962.

99. Smulevich, G.; Spiro, T. G. *J. Phys. Chem.* **1985**, *89*, 5168.

100. Hildebrandt, P.; Stockburger, M. *J. Phys. Chem.* **1986**, *90*, 6017.

101. Paul, K.-G. *Acta Chem. Scand.* **1950**, *4*, 239.

102. Siiman, O.; Bumm, L. A.; Callaghan, R.; Blatchford, C. G.; Kerker, M. *J. Phys. Chem.* **1983**, *87*, 1014.

103. Hildebrandt, P.; Stockburger, M. *Biochemistry* **1989**, *28*, 6710.

104. Cotton, T. M.; Schelegel, V. L.; Holt, R. E.; Swanson, B.; de Montellano, P. O. In *Raman Scattering Luminescence and Spectroscopic Instrumentation in Technology*, SPIE Vol. 1055: Bellingham, WA, 1989; p. 263.

105. de Groot, J.; Hester, R. E. *J. Phys. Chem.* **1987**, *91*, 1693.

106. de Groot, J.; Hester, R. E.; Kaminaka, S.; Kitagawa, T. *J. Phys. Chem.* **1988**, *92*, 2044.

107. Niki, K.; Kawasaki, Y.; Kimura, Y.; Higuchi, Y.; Yasuoka, N. *Lamgnuir* **1987**, *3*, 982.

108. Cotton, T. M.; Schlegel, V. L.; Rospendowski, B. N.; Uphaus, R. A.; Wang, D. L.; Eng, L. H.; Stankovich, M. T. In *Raman Scattering Luminescence and Spectroscopic Instrumentation in Technology*, SPIE, in press.

109. Cotton, T. M.; Timkovich, R.; Cork, M. S. *FEBS Lett.* **1981**, *133*, 39.

110. Lewis, D. F. V. *Drug Metabol. Rev.* **1986**, *17*, 1.

111. Champion, P. M.; Gunsalus, I. C.; Wagner, G. C. *J. Am. Chem. Soc.* **1978**, *100*, 3743.

112. Kelly, K.; Rospendowski, B.; Smith, W. E.; Wolf, C. R. *FEBS Lett.* **1987**, *222*, 120.

113. Wolf, C. R.; Miles, J. S.; Seilman, S.; Burke, M. D.; Rospendowski, B. R.; Kelly, K.; Smith, W. E. *Biochemistry* **1988**, *27*, 1597.

114. Hildebrandt, P.; Greinert, R.; Stier, A.; Stockburger, M.; Taniguchi, H. *FEBS Lett.* **1988**, *227*, 76.

115. Rospendowski, B. N.; Schlegel, V. L.; Holt, R. E.; Cotton, T. M. In *Charge and Field Effects in Biosystems-2*, Allen, M. J.; Cleary, S. F.; Hawkridge, F. M. Eds.; Plenum Press: New York, 1989; p. 43.

116. Ni, F.; Cotton, T. M. *J. Raman Spoectrosc.* **1988**, *19*, 429.

117. Thomas, E. W.; Merlin, J. C. *Spectrochim. Acta* **1979**, *35A*, 1251.

118. Terada, H.; Kim, B.-K.; Saito, Y.; Machida, K. *Spectrochim. Acta* **1975**, *31A*, 945.

119. Smulevich, G.; Feis, A. *J. Phys. Chem.* **1986**, *90*, 6389.

120. Smulevich, G.; Feis, A.; Mantini, A. R.; Marzocchi, M. P. *Indian J. Pure Appl. Phys.* **1988**, *26*, 207.

121. Oesterhelt, D.; Stoecckenius, W. *Nature* **1971**, *223*, 149.

122. Nabiev, I. R.; Efremov, R. G.; Chumanov, G. D. Proc. Int. Symp. Biomol. Struct. Interactions *Suppl. J. Biosci.* **1985**, *8*, 363.

123. Nabiev, I. R.; Chumanov, G. D.; Efremov, R. G. *J. Raman Spec.* **1990**, *21*, 49.

124. Cooney, R. P.; Mahoney, M. R.; Howard, M. W. *Chem. Phys. Lett.* **1980**, *76*, 488.

125. Lutz, M.; Agalidis, I.; Hervo, G.; Cogdell, R. J.; Reiss-Husson, F. *Biochim. Biophys. Acta* **1978**, *503*, 287.

126. Cotton, T. M.; Van Duyne, R. P. *FEBS Lett.* **1982**, *147*, 81.

127. Feher, G.; Okamura, M. Y. In *The Photosynthetic Bacteria*, Clayton, R. K.; Sistrom, W.R. Eds.; Plenum Press: New York, 1978, and references cited therein.
128. Picorel, R.; Holt, R. E.; Cotton, T. M.; Seibert, M. In *Progress in Photosynthesis Research*, Biggins, J. Ed.; Martinus Nijhoff: Dordrecht, 1987, Vol. 1, p. I.4.423.
129. Picorel, R.; Holt, R. E.; Cotton, T. M.; Seibert, M. *J. Biol. Chem.* **1988**, *263*, 4374.
130. Picorel, R.; Lu, T.; Holt, R. E.; Cotton, T. M.; Seibert, M. *Biochemistry* **1990**, *29*, 707.
131. Brunisholz, R. A.; Zuber, H.; Valentine, J.; Lindsay, J. G.; Wolley, K. J.; Cogdell, R. J. *Biochem. Biophys. Acta* **1986**, *849*, 295.
132. Webster, G. D.; Cogdell, R. J.; Lindsay, J. G. *FEBS Lett.* **1980**, *111*, 391.

THREE-DIMENSIONAL CONFORMATIONS OF COMPLEX CARBOHYDRATES

C. Allen Bush and Perseveranda Cagas

Advances in Biophysical Chemistry, Volume 2, pages 149–180
Copyright © 1992 by JAI Press Inc.
All rights of reproduction in any form reserved.
ISBN: 1-55938-396-8

I. INTRODUCTION: WHAT ARE COMPLEX CARBOHYDRATES?

A. Definition

Complex carbohydrates could be defined in as many ways as there are research groups in the field, but for the purposes of this discussion we choose to consider polymers or oligomers composed of such monosaccharides as mannose, galactose, *N*-acetyl glucosamine, *N*-acetyl galactosamine, rhamnose, fucose and sialic acid connected in highly varied linkages and anomeric configurations. Typically they vary in size from disaccharides to oligosaccharides of 15 to 20 residues in length or they may occur as polysaccharides of 100,000 daltons composed of repeating subunits containing up to 8 sugars. Since there are multiple points of possible linkage on a monosaccharide, branching of the chain is not only possible, it is in fact quite common in complex carbohydrates.

It is somewhat difficult to define complex carbohydrates without reference to the specific macromolecules in which they occur, most commonly in a covalent linkage with other biomolecules such as lipids or proteins. The complex carbohydrates on which attention will focus in this review are found in glycoproteins, glycolipids and in the extracellular polysaccharides of bacteria. Closely related complex carbohydrates include the complex glycans of glycosyl phosphatidyl inositol anchors of membrane proteins and proteoglycans such as heparin, chondroitin and dermatan. In nature, the proteoglycans are covalently linked to protein and thus could also be considered as glycoproteins but the complex carbohydrates of the glycosaminoglycans, including hyaluronic acid are structurally distinct and will not be discussed in detail in this review. One common thread among all these complex carbohydrates is that they occur in rather small proportions in the cells of higher organisms, a feature which makes study by chemical methods difficult. A second common property of the complex carbohydrates is that none of them has a purely structural role or serve as energy sources for the organism. Thus starch, glycogen, chitin and cellulose can all be distinguished from complex carbohydrates on functional grounds.

B. Function of Complex Carbohydrates

While the role of cellulose in cell wall structure or of glycogen in energy storage can be stated with some precision, a similar formulation of the function of complex carbohydrates cannot be made so readily. In general terms, it is thought that complex carbohydrates serve a role in cellular signaling but examples of specific biological function are very few. In the absence of any unifying generalization concerning the function of complex carbohydrates, we must summarize a few specific facts from which we can infer a vague idea of function. Although the biological function of the complex N-asparagine-linked carbohydrates of glycoproteins remains unknown, the considerable data on their biosynthesis accumulated in recent years points to a remarkably complicated scheme of specific anabolism, followed by an equally specific pathway of programmed catabolism and resynthesis.[1-3] This complex scheme, which is conserved in the golgi apparatus of the cells of eucaryotic species as diverse as yeast and man, consumes too much biochemical free energy to be without an important biological function. The O-linked oligosaccharides of glycoproteins differ in many details from their N-linked counterparts but are equally complex and varied in structure as are the glycolipid carbohydrate structures found in membranes of eucaryotic cells. Biological data point to a function for glycoproteins in the control of growth and differentiation in the complex cell organization of eucaryotic systems and some complex oligosaccharides appear to act as tumor antigens.[4] Carbohydrate structures related to blood group oligosaccharides are implicated in the recruitment of neutrophils at the site of inflammation by binding of endothelial leukocyte adhesion molecule.[5]

The general argument made above for the role of complex carbohydrates as informational macromolecules implies that there must be a mechanism for decoding the information which is stored in their structures. Apparently the information is decoded by a group of multivalent carbohydrate-binding proteins known as lectins.[6] A wide variety of these proteins have been found in animal tissues.[7] Some lectins are membrane bound and some are soluble and both are thought to have important biological functions.

Although it is the animal lectins which are the important targets for biochemical study, their low abundance in animal tissues will present serious obstacles and it is likely that advances in this research will require the methods of molecular biology. The lectins isolated from plants have served as valuable models for understanding carbohydrate binding since they are available in quantity. But even these systems have demonstrated daunting complexity due to multivalency. It is generally found that binding of a single carbohydrate determinant is relatively weak, with $K_d \approx 10^{-4}$ molar. But modifications of cell function by in vivo lectin binding show activity at much lower concentrations. Since lectins are always found as active protomers of at least two subunits, their binding to multiple carbohydrates on a cell surface raises the effective binding constant. Such cooperative interactions in which a number of weak effects act in concert to give a strong interaction are not rare in biochemical phenomena. Not only does lectin activity generally involve binding of multiple carbohydrate sites to a single lectin molecule, but multiple sites on a single carbohydrate chain can also bind two or more lectin sites if the geometric placement of the lectin binding determinants is suitable.[8] This observation raises the possibility for a third type of cooperative interaction, that between the lectin molecules. In fact, electron microscope evidence exists for the formation of exactly such a cooperative complex of plant lectins and the multivalent carbohydrate chains of N-linked glycopeptides.[9] To further complicate study of the interaction between complex carbohydrates and lectins, the formation of lectin oligomers depends on both pH and on bivalent metal ions typically Mg^{++} or Ca^{++}. In a system of such complexity, it is not surprising that a detailed understanding has not yet been achieved. Very careful studies by the techniques of biophysical chemistry along with some exceptionally thoughtful analysis will be required before we can feel confident about the interpretation of biologically relevant interactions between multivalent lectins and complex carbohydrates.

While the surfaces of animal cells are covered with complex carbohydrate structures of glycoproteins, glycolipids and proteoglycans, the surfaces of bacterial cells are devoid of these structures which are unique to eucaryotes. Yet the surfaces of bacteria are also decorated with sugars in the form of polysaccharides composed of repeating subunits which are remarkably similar in their chemical makeup to the complex carbohydrates of eucaryotic cells. Most of the complex polysaccharides of the bacterial surface occur in either of two covalent complexes. The first class is the O-antigens of the lipopolysaccharides found on gram negative bacteria. These polysaccharides, generally composed of repeating subunits, are attached to a carbohydrate core substituted by fatty acid chains called lipid A which is inserted in the bacterial membrane to anchor the polysaccharide. Bacterial polysaccharides can also be covalently attached to the cell wall, which is itself composed of polysaccharide which is crosslinked by a peptide. Under certain circumstances, these complex polysaccharides can be shed from the cell in the form of capsular polysaccharides or exo-polysaccharides which are often recovered from the culture medium. The identities of the constituent monosaccharides and the biosynthesis of these bacterial polysaccharides are generally distinct from those of mammalian glycoproteins and glycolipids. The former are more highly varied in carbohydrate composition, anomeric form and glycosidic linkage but there are nevertheless some similarities in the chemical structures and many cross reactions of lectins and antibodies between bacterial surface polysaccharides and the cell surfaces of higher organisms have been observed. The bacterial cell surface polysaccharides act as receptors for many plant lectins as well as for fimbrial lectins found on the surface of other bacterial cells. These bacterial coaggregation interactions contribute to bacterial adhesion and play an important role in bacterial ecology.

The interactions of the lectins and the complex polysaccharides of the bacterial surface with the eucaryotic cell surface have some important practical consequences. Both bacterial surface polysaccharides and lectins afford attachment points for colonization and pathogenesis of an animal host organism, allowing the bacteria to flourish in the hostile environment presented by the normal defense mechanisms of the host tissues. An impressive demonstration of biologically significant cross reactions between the structures of glycoproteins, glycolipids and bacterial polysaccharides is demonstrated by the following experiment. A mixture of glycolipids isolated from mammalian tissue having a wide variety of complex carbohydrate chains is fractionated on a silica thin-layer plate. The plate is then overlaid with radioactively labeled bacteria known to colonize and be pathogenic for that tissue. Adherence of the bacteria to the glycolipid oligosaccharides mediated by bacterial surface lectins can be readily detected and many highly specific interactions have been found by this method which suggest a role for these lectin-oligosaccharide interactions in the pathogenicity of bacteria in these tissues.[10,11]

The relationship between the capsular polysaccharides and the *O*-antigens of the lipopolysaccharides and the pathogenicity of bacteria has been exploited for diagnostic purposes in the field of medical microbiology for many years. A possible interpretation of the role of these polysaccharides in pathogenicity is that the bacteria have evolved lectins and polysaccharides which are complementary to the glycoproteins, glycolipids and cell surface lectins of mammalian tissues to enable them to evade the host defenses and to provide a suitable environment for colonization of the host tissues. While we do not yet understand the origin of the animal's requirement for these glycoproteins, glycolipids and associated lectins, the pathogenic bacteria seem to understand this need and have evolved a mechanism to exploit it.

Both the chemical similarities and the biological cross reactions suggest that studies of bacterial polysaccharides in concert with glycoproteins or glycolipids of higher organisms could offer a profitable route to valuable insights into the latter. Studies of the carbohydrate structure of glycoproteins present numerous difficulties resulting from their small quantities in nature and their substantial heterogeneity which must be resolved by high resolution chromatography. Bacterial polysaccharides, in contrast, can be much more agreeable targets for study by the methods of biophysical chemistry such as NMR spectrometry and X-ray crystallography. The organisms are generally easy to grow in quantity and isolation of pure polysaccharides from bacterial sources is generally facilitated by the simplicity of the biological system which contains many fewer gene products than does a eucaryotic cell. Not only is isolation of substantial quantities of pure sample facilitated, but it is also possible to quantitatively incorporate stable isotopes in bacterial polysaccharides, a feat which is almost impossible in complex oligosaccharides isolated from mammalian sources. The recently introduced heteronuclear NMR methods (specifically ^{1}H detected ^{13}C spectroscopy) are extremely powerful when applied to problems of structure determination of proteins.[12] The problem of determination of the conformation of the side chains of the amino acids in proteins presents many of the same problems of NMR signal resolution which are present in complex carbohydrates.[13] The spectacular successes of 3-dimensional and 4-dimensional NMR spectroscopy with ^{13}C enriched proteins suggests that the same approach might meet with some success in complex carbohydrates.

II. CONFORMATION OF COMPLEX CARBOHYDRATES

A. General Consideration of the Magnitude of the Problem

A detailed understanding of the interaction of lectins with their complex carbohydrate receptors requires study of the conformation of both partners. The conformation of proteins in general and of plant lectins in particular has been very successfully investigated, primarily by means of X-ray crystallography, a subject which is reviewed in the pages of this volume by Howard and Poulos. The complete three-dimensional structures of several plant lectins have been reported already, some including a bound carbohydrate.[14,15] The substantial homology in amino acid sequence among many of the plant lectins offers the promise that some useful generalizations can be made about the conformation of the binding sites of these proteins.[16] It is also possible that the combination of NMR spectroscopy and molecular modeling methods, which have been applied to smaller proteins, may prove useful in the study of peptide conformation of the sugar binding sites of lectins. In contrast to the progress of biophysical chemists in protein three-dimensional structure, stands the rather primitive understanding of the conformation of the complex carbohydrate. Although X-ray crystallography could be applied to complex carbohydrates, only a very few structures of carbohydrates from glycoproteins, glycolipids and bacterial polysaccharides have been reported. Perhaps the scarcity of X-ray structures stems less from any fault of the crystallographic method than from the dearth of sufficiently pure samples of complex carbohydrates suitable for crystallization. Much greater progress in the conformation of complex carbohydrates has been achieved by use of NMR spectroscopy and molecular modeling. This approach requires a smaller sample of the target compound and the requirements for purity are less stringent than are those for crystallography. Therefore it is these latter methods which will be emphasized in this review.

In order to determine the conformation of the carbohydrate determinant the number of sugar residues making up a binding site of a lectin must be known. Carbohydrate binding data on plant lectins imply that while some of them bind only a single sugar residue, others seem to bind at least a disaccharide. Although there is little evidence in the literature for any plant lectin with a binding site which accommodates anything larger than a disaccharide, we do not know whether animal lectins might bind larger oligosaccharides. One group of biologically important carbohydrate determinants are the blood group oligosaccharides whose structures will be discussed below. These complex oligosaccharide determinants are composed of three to four sugar residues. Since the true biological function of blood group activity is unknown, we do not know what type of lectin is the natural receptor for these structures. An upper limit to the size of a lectin binding site of about six monosaccharide residues can be determined by the size of a typical globular protein, a classification into which all known lectins fall. Such a site is that for the binding of the hexasaccharide substrate for lysozyme which stretches across the cleft of this small protein. Therefore, the problem of the conformation of the oligosaccharide binding site of a lectin can be focused on small oligosaccharides of approximately 2 to 4 or perhaps as many as 6 residues. We must ask whether oligosaccharides of this size adopt rigid structures or whether they are flexible. This is an important issue since any internal motion must be frozen out in the binding of the oligosaccharide to the lectin receptor.

It will be recognized that complex oligosaccharides such as those of glycoproteins, glycolipids and bacterial polysaccharides may contain many more than 1 to 4 residues. For example, the asparagine *N*-linked glycopeptides commonly have as many as 15 residues. Thus they may be multivalent and there is some evidence that a single glycopeptide can indeed bind at two or more lectin sites in a cooperative interaction. Thus the conformation and dynamics of the sugars spanning the two lectin binding sites becomes relevant to any interaction. If there is flexibility of the link between the two binding sites, then those degrees of freedom will be frozen out by the cooperative binding and will be quite important to the overall free energy of the interaction. Therefore the problem of the conformation of complex carbohydrates can be divided into two parts, the first being the local interaction over the range of two to four monosaccharides and the second being the overall conformation, specifically that of the branch points joining two arms each of which might carry the same or different lectin binding determinants.

B. NMR Methods for Studying Conformation of Complex Oligosaccharides

The conformation of complex carbohydrates has been most successfully approached by a combination of NMR spectroscopy and molecular modeling following technology which has been developed for the study of peptides and proteins. This study will not attempt to review the truly elegant developments in this exciting field of protein structure determination in solution. The topic has been summarized by the research groups responsible for development of the technology.[17–20] This review will emphasize those specific features of NMR spectroscopy and molecular modeling which are unique to complex carbohydrate chemistry. In some ways, the NMR spectroscopy and molecular modeling of complex carbohydrates is more difficult than that of proteins but there are many ways in which it is easier.

The first step in application of the method, regardless of the target molecule, is assignment of the ^1H NMR spectrum. That is, one must establish the correspon-

$[\rightarrow 6)Gal_pNAc\ \alpha(1\rightarrow 3)Rha_p\ \beta\text{-}(1\rightarrow 4)Glc_p\ \beta\text{-}(1\rightarrow 6)Gal_f\ \beta\text{-}(1\rightarrow 6)$

$Rha\ \alpha\text{-}(1\rightarrow 2)$ ⟋ $\text{-}Gal_pNAc\ \beta\text{-}(1\rightarrow 3)Gal_p\ \alpha\text{-}(1\rightarrow PO_4^-\ -]_n$

Scheme 1.

dence between all the protons of the molecule and the resonances in the spectrum. This is accomplished by determining the spin coupling correlation which generally exists between vicinal (or geminal) protons in the structure. The procedure for complex carbohydrates is essentially identical to that for peptides and proteins. The coherence through two-bond or three-bond coupling was detected in earlier work by 1-dimensional difference decoupling[21] but the method of 2-dimensional phase sensitive COSY spectroscopy has largely replaced this method because it is easier and faster.[22] Figure 1 shows the double-quantum filtered phase-sensitive COSY spectrum in D_2O solution of a bacterial polysaccharide from *Streptococcus mitis* strain J22 which is composed of seven monosaccharides in repeating subunit the structure of which is shown in Scheme 1. Although this polymer has a molecular weight of approximately 100 kD, it gives a much better resolved ^1H NMR spectrum than would a protein of similar molecular weight as a result of internal motion in the polymer. Thus its effective T_2 is comparable to that in a small protein and COSY cross peaks are readily detected. Since each subunit in the polymer is chemically equivalent, the spectrum has the appearance of that of a heptasaccharide without a reducing terminal sugar.

Assignment of this spectrum by coupling correlation requires an initial point for identification of the individual spin systems or sugar rings. Since the anomeric proton is connected to a carbon bearing two oxygen atoms, it is generally the most downfield of the ^1H signals and makes a convenient starting point for the assignment. The seven anomeric proton resonances in Figures 1 and 2 fall between 5.4 and 4.4 ppm. With a generally well resolved chemical shift, the anomeric proton resonance plays a role somewhat similar to that of the amide proton resonance in the scheme for assignment of the resonances of peptides. The spectroscopic assignment of a sugar follows the chain of spins just as in an amino acid side chain. A second NMR technique which is a useful adjunct in tracing out the spin system is the isotropic mixing experiment known as HOHAHA or TOCSY. In Figure 2, showing the 2-d HOHAHA spectrum of this same polysaccharide, cross peaks represent the relay of magnetization through the entire spin system resulting in cross peaks between all the resonances in any given monosaccharide residue. This experiment is especially useful in sugars, for which the similar chemical shifts of the methine protons lead to many examples of strong coupling or intermediate coupling. In these cases, two correlated signals are close in the spectrum causing the COSY crosspeak to be close to the diagonal where it may be impossible to

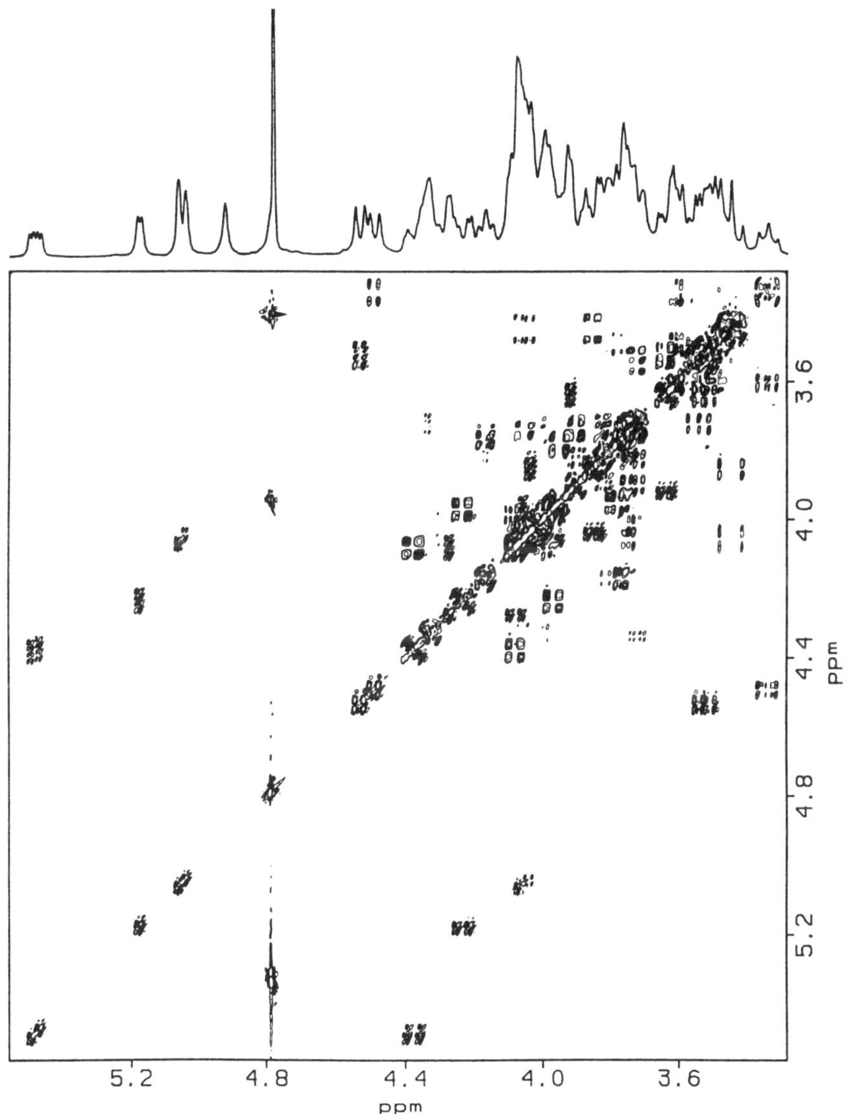

Figure 1. 300 MHz double quantum filtered COSY spectrum of the polysaccharide from *S. sanguis* J22 in D_2O, 24°, taken from Ref. 23 with permission.

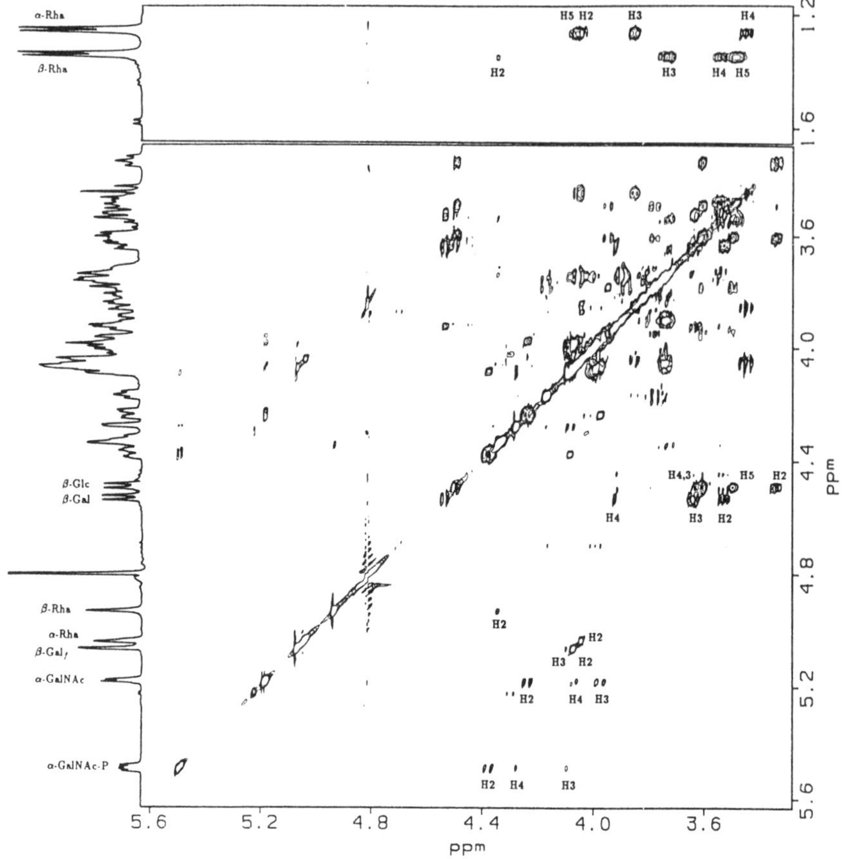

Figure 2. 500 MHz 2D HOHAHA spectrum of the polysaccharide from *S. sanguis* J22 in D₂O, 24°, taken from Ref. 23 with permission.

detect. But in the 2-d HOHAHA spectrum, which contains cross peaks between any isolated resonance such as H1 and the other resonances in the spin system, the relevant peaks will be well removed from the diagonal.

The anomeric proton resonance is not the only possible "reporter group" signal useful for initiating identification of a spin system by COSY or HOHAHA. Since the resonance of the methyl group of the 6-deoxy hexose, rhamnose, occurs well upfield (at 1.3 ppm) in Figure 2, those cross peaks can serve as an alternative starting point for spin system assignment. Other examples of 6-deoxy hexoses in addition to rhamnose commonly encountered in complex carbohydrates include fucose and quinovosamine. Sialic acids and ketodeoxyoctanoate (KDO), both of which occur with a ketosidic linkage, lack an anomeric proton but the methylene protons of these 3-deoxy sugars have distinctive chemical shifts near 2.5 and 1.8 ppm which can serve as signature resonances for the ring spin system assignment.

An additional method for defining a signature resonance for anchoring an assignment is triple-quantum-filtered COSY which specifically detects systems of three mutually coupled spins having distinct chemical shifts. The most common occurrence of such systems in complex carbohydrates is in the H6, H6' and H5 of hexoses. The utility of the TQF-COSY experiment is greatest for such monosaccharides as galactose, in which very small coupling constant values correlate H4 with H5, preventing a complete spin analysis by HOHAHA.[23] In summary, the main obstacles to complete spin system identification arise from strong coupling and small values of $^3J_{HH}$.[22] In the most troublesome cases, homonuclear ^1H methods cannot provide an unambiguous assignment and the use of ^{13}C resolved spectroscopy is required. The topic of 2-dimensional ^1H-detected ^{13}C spectroscopy (HMQC) will not be covered in this review but it has been shown that natural abundance ^{13}C is very effective in spreading out the ^1H spectrum in the ^{13}C dimension, greatly improving the resolution and eliminating the effects of strong ^1H coupling.[23] For extreme cases, combination methods such as HMQC-COSY can be used to generate cross peaks between not only the ^1H resonance and that of its bound carbon atom at its characteristic ^{13}C chemical shift but also COSY crosspeaks at the ^1H frequency of homonuclear coupled protons.[24] Ultimately these hybrid methods can be extended to three dimensions, in which a proton homonuclear 2-d spectrum is further resolved by the ^{13}C chemical shift in the third dimension.[25] There are many parallels between the assignment problem in protein side chains and in complex carbohydrates. The carbohydrate chemical shift dispersion is smaller than that in peptides due to the lack of aromatic groups in sugars. The resulting overlap in chemical shifts is not unlike that found in the aliphatic amino acid side chains.

Once the ^1H assignment is complete, the next step in the determination of conformation is measurement and interpretation of the nuclear Overhauser effects (NOE). In this experiment, the contributions of all the protons to the relaxation of the magnetization of a given proton through the dipolar relaxation mechanism are measured. In the earliest work on oligosaccharides, this measurement was done by the steady state NOE method in which a proton resonance is selectively irradiated during the preparation period followed by quantitative measurement of the change in signal intensity of all other peaks in the spectrum by the 1-dimensional difference spectroscopy method.[26–29] This method has been largely replaced by phase-sensitive 2-d NOESY methods which have several technical advantages. Not only does the 1-d difference method require exceptional spectrometer stability, but it also requires substantial interaction of the operator with the instrument which does not make optimal use of spectrometer time. A NOESY spectrum acquired in a single run of 4 to 16 hours provides all the relevant data without operator intervention and the data analysis can be carried out separately on a computer workstation. Also, quantitative interpretation of 1-d difference data is complicated by the overlapping of resonances so common in complex carbohydrates. Specifically, it is impossible to selectively irradiate signals which resonate within about 20 to 80 Hz since a selective saturating or inverting pulse is in practice about 50–100 Hz wide. The added resolution introduced by the second dimension in the NOESY experiment often provides quantitative NOE data for such closely spaced peaks increasing the number of the data useful in defining the glycosidic torsion angles, Ψ and Φ.[68]

Since the proton-proton dipolar cross relaxation which leads to the NOESY crosspeaks depends on the inverse sixth power of the distance between the protons, r_{ij}^{-6}, the size of the crosspeaks is a measure of interproton distances and provides information which may be used in determining oligosaccharide conformation. We will discuss in more detail below methods for combining molecular modeling and a quantitative interpretation of the buildup and decay kinetics of NOESY cross peaks to provide rigorous analysis of the data in terms of geometric models.

Another source of information in the ^{1}H NMR data comes from the vicinal coupling constants which are related to dihedral angles. Although homonuclear ^{1}H—^{1}H coupling constants can be measured from either 1-d or 2-d spectra, the poor digital resolution of the latter may pose a technical problem if highly accurate values of $^3J_{HH}$ are required. In a pyranoside, it is found that the six-membered ring generally forms a chair of fixed conformation providing a classification of protons as axial or equatorial and the coupling patterns are characteristic of the stereochemistry of the monosaccharide residue.[22] For example, if H2 is axial, as it is for the gluco and galacto stereochemistry, then a small coupling constant $^3J_{H1-H2}$ of 2–3 Hz is observed for α-anomers as a result of the gauche conformation of H1 and H2 following the Karplus relation. The *trans* diaxial relationship of H1 and H2 in β-anomers of sugars with the gluco and galacto configuration leads to larger coupling constants of 7–9 Hz for $^3J_{H1-H2}$. A concise rule for pyranosides is that axial-axial coupling constants are 7–11 Hz and any couplings involving equatorial protons range from 1–3 Hz. In furanosides, these simple rules do not operate due to the more complicated puckering of five-membered rings.

In addition to the sugar ring pucker derived from the $^3J_{HH}$ it may be possible to extract information from coupling constants of exocyclic groups. The vicinal coupling constants between H5 and H6 report on the dihedral angle about the C5—C6 bond a parameter which is crucial in the case of (1→6) linkages. Interpretation of $^3J_{H5-H6}$ generally requires stereospecific assignment of the two H6 protons. On the most important dihedral angles, Φ and Ψ describing the glycosidic linkage, $^3J_{H-H}$ cannot provide any information but the heteronuclear long range coupling constants, $^3J_{C-H}$ could in principle be very useful.[30–31] Unfortunately it is technically difficult to measure these long range couplings in complex carbohydrates in natural abundance ^{13}C and this method has not yet provided much useful information relevant to the problem.

The experimental methods summarized above provide some fragmentary information on the conformation of an oligosaccharide. NOESY data can be quantitatively interpreted by the theory of nuclear relaxation. The coupling constants can be interpreted by trigonometric formulas with suitable parameters derived from empirical studies on model compounds. But the distances and dihedral angles thus derived are inadequate to uniquely determine the complete conformation of a complex carbohydrate. In fact, the known covalent structure of the oligosaccharide can provide sufficient additional data so that the method of computer-based molecular modeling can be combined with the NMR data to yield a detailed geometric model of the oligosaccharide. The molecule is simulated by a classical potential or force field which describes the interaction of the component atoms following methods which have been widely used in peptides and proteins. The subject of carbohydrate molecular modeling has been reviewed in an earlier volume in this series[32] and relevant aspects of this topic will be touched on in the examples below.

III. NOE STUDIES OF BLOOD GROUP OLIGOSACCHARIDES

A. Proposal for Relatively Rigid Oligosaccharides

The conformations of blood group oligosaccharides were first studied by Lemieux et al.[33] who reported carbon and proton NMR spectroscopic assignments for several trisaccharide fragments which model the non-reducing terminals of blood group A,B, H and Lewis oligosaccharides. They also reported qualitative NOE data along with conformational energy calculations employing a method which assumes that only non-bonded energies and the torsional potential about one of the glycosidic dihedral angles (exo-anomeric effect) are important. In these calculations which ignore hydrophobic effects, hydrogen bonding and other electrostatic interactions, a simple Kitaigorodsky form was used for the non-bonded interactions with the parameter set reported by Venkatachalam and Ramachandran.[34] This formulation of the non-bonded interactions had been used in previous studies of carbohydrate conformation by Rao and coworkers.[35] On the basis of these results, Lemieux et al.[33] proposed single conformations for the blood group oligosaccharides which are determined mainly by steric or non-bonded interactions and by torsional potentials. The influence of hydrogen bonding and other electrostatic effects and of hydrophobic bonding were proposed to be unimportant. The experimental NMR data and the method of interpretation as well as the molecular modeling are all open to criticism. Although the conclusions rely on a number of poorly tested assumptions, they are very instructive and can be taken as a working hypothesis which is subject to test by other methods.

The next step was the study of related compounds with an attempt to apply more rigorous tests to these proposals by Lemieux and coworkers to more precisely determine their validity and generality. In experimental studies, we have used various oligosaccharides isolated from natural sources which contain the blood

Table 1. Structures of Blood Group Oligosaccharides

Oligosaccharide	Primary Structure
R6 (Blood Group A)	Fuc α-(1→2)⟍ Gal β-(1→3)GalNAc-ol GalNac α-(1→3)⟋
LNF-1	Fuc α-(1→2)Gal β-(1→3)GlcNAc β-(1→3)Gal β-(1→4)Glc
LNF-2 (Lewis[a])	Fuc α-(1→4)⟍ GlcNAc β-(1→3)Gal β-(1→4)Glc Galβ(1→3)⟋
R9 (Blood Group A)	Fucα-(1→2)⟍ Galβ-(1→3)GlcNAc β-(1→3)GalNAc-ol GalNAc α-(1→3)⟋
R5 (Blood Group A)	Gal β-(1→3)⟍ Fuc α-(1→2)⟍ GalNAc-ol Gal β(1→4)GlcNAc β-(1→6)⟋ GalNAc α-(1→3)⟋
Difucohexasaccharide (Blood Group H)	Fuc α–(1→2)Gal β-(1→3) ⟍ GalNAc-ol Fuc α-(1→2)Gal β-(1→4)GlcNAc β-(1→6)⟋
LND-1 (Lewis[b])	Fuc α-(1→4)⟍ GlcNAc β-(1→3)Gal β-(1→4)Glc Fuc α-(1→2)Gal β-(1→3)⟋

group A, H and also Lewis blood group determinants the structures of which are shown in Table 1.

B.   Quantitative Interpretation of NOE Data

The outline of the experimental NMR approach to this problem has been given above. After complete assignment of the ¹H signals, quantitative NOE are recorded. It should be mentioned that a technical problem encountered in obtaining high quality NOE data for oligosaccharides in the general size range of those in Table 1 results from unfortunate values of the rotational correlation times of many oligosaccharides containing from four to six sugar residues. For such cases, the rotational correlation time may be close to the reciprocal of the spectrometer frequency causing the observed NOE to be nearly zero as a result of the approximate cancellation of the spin-lattice relaxation with the effect of cross relaxation. In our research we have circumvented this problem controlling the rotational correlation time of the oligosaccharide by modification of the probe temperature and the solvent viscosity.[36,37]

The NOE data may be obtained by the steady-state 1-d difference method or preferably by the phase-sensitive 2-d NOESY method. A typical NOESY spectrum shown in Figure 3 contains a number of cross peaks indicating proximity both for protons within a sugar ring and also between protons on different residues.

Although some workers have used a simple method for interpretation of the NOESY cross peak data in which ratios of cross peaks are taken as the ratios of the sixth power of the interproton distances, there are some possible criticisms of this isolated spin pair approximation (ISPA). For the interpretation of the NOE data on complex carbohydrates, we have advocated a complete spin matrix approach in which mutual relaxation pathways among all the protons on the oligosaccharide are considered simultaneously. There are three reasons for preferring this method over the more commonly used ISPA method. The first reason stems from the abundance of closely spaced protons in sugars which makes it difficult to define isolated spin pairs and there are numerous "three spin effects" found in the notation of Noggle and Schirmer.[38] A second reason arises from the unfavorable rotational correlation time of oligosaccharides of the size shown in Table 1 which leads to generally smaller values for the NOE than are found for most small proteins. Thus the requirement of the ISPA method for measuring accurate NOESY data at short mixing times where the NOE buildup curve is linear is technically difficult. A complete spin matrix method allows one to quantitatively interpret NOESY data at any value of the mixing time. A third argument for the complete spin matrix method is the observation of a number of very small yet very important NOESY peaks at about the 2–5% level in complex oligosaccharides. In Figure 3, there appear several small crosspeaks between protons on residues which are not directly connected and correct interpretation of these data provide information on two or more glycosidic linkages simultaneously. It is difficult in the ISPA method to distinguish whether these small long range NOE arise from direct relaxation pathways or whether they are indirectly relayed three spin effects. Our results on blood group oligosaccharides demonstrate that these small effects are extremely valuable in distinguishing rigid from flexible structures and in establishing precise conformations.

NOE data are simulated by assuming a model conformation, calculating the NOE spectrum predicted for that conformation and then adjusting the geometry of the model in an attempt to match the experimental data. For simulation of steady state NOE, the elements of the relaxation matrix, ρ_{ii} and σ_{ij}, are calculated from the geometric model in which a single isotropic rotational correlation time, τ_c, is assumed.[38] The value of this latter parameter can either be estimated from NOE data within a single sugar ring whose values do not depend strongly on geometry or they may be calculated from $^{13}C\ T_1$ data. The steady state NOE can then be calculated by a matrix inversion method.[26–28,39] Simulation of transient NOE data

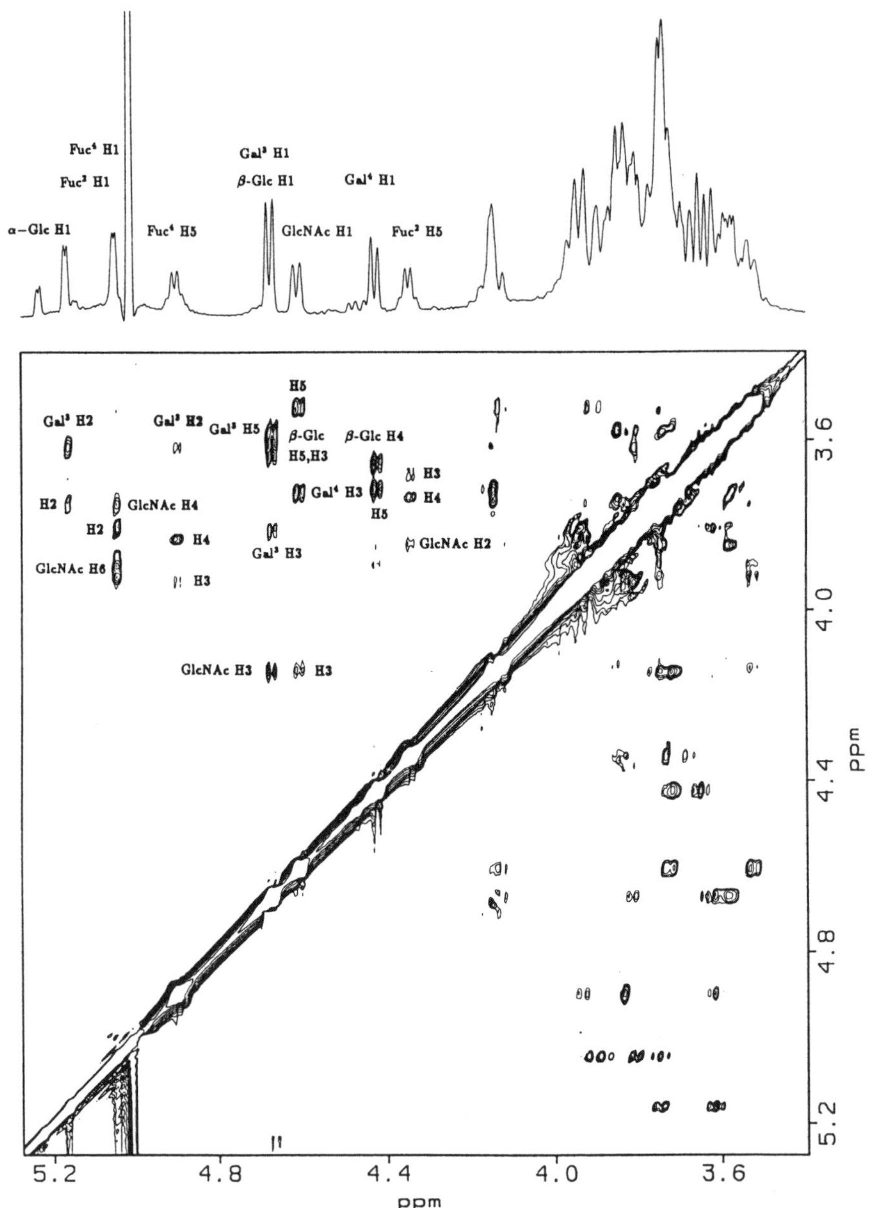

Figure 3. 500 MHz NOESY spectrum of LND-1 in D_2O, t_{mix} = 250 ms, 5°, taken from Ref. 37 with permission.

164

such as those from a 2-d NOESY experiment requires the same input data and a similar calculation.[40] The relaxation matrix is diagonalized and NOESY cross peaks can be calculated as function of mixing time. This calculation yields the ratio of the crosspeak volume to the volume of a diagonal peak from a single proton at zero mixing time, $t_m = 0$. Simulation of a NOESY spectrum at a single mixing time is possible if selective T_1's are known since the decay of the diagonal peak with mixing time is the same as when that peak is selectively inverted.[41-43] The details of the application of this method for quantitative simulation of NOESY crosspeaks at a single mixing time were recently described.[37]

The matrix method outlined here is effectively similar to the BACKCALC method which has been used in peptides and nucleic acids.[44] Both methods implicitly assume isotropic motion at a single rotational correlation time but this assumption can be partially relaxed by using adjustable values of τ_c in the matrix method or by using different values of the "Z-leakage" rate in the method described by Summers et al.[44] The latter method requires measurement of NOESY buildup curves while the matrix method substitutes selective T_1 measurements to obtain the kinetics.

C. Application to Blood Group Oligosaccharides

In order to apply this method we must first assume a single rigid isotropically tumbling conformation for the oligosaccharide. Although this assumption is a significant one which must be validated, it is subject to test within this method. For oligosaccharides composed of pyranosides we assume rigid chairs, an assumption supported both by $^3J_{H-H}$ data and by molecular dynamics simulations.[32] Under this assumption, the conformation of a disaccharide can be specified by two dihedral angles Φ and Ψ which describe rotations about the bonds to the glycosidic oxygen atom. For a $(1\rightarrow6)$ linkage a third dihedral angle, ω, is required to describe rotation about the C5—C6 bond. The complete NOE matrix for a disaccharide may then be calculated as a function of these dihedral angles in some modest angle increment such as 10° and the results matched with the experimental NOE data. The results shown in Figure 4 appear somewhat similar to energy maps but they represent those parts of the conformational space in which experimental NOE ratios agree with those calculated from the model. For a trisaccharide or a tetrasaccharide, the search over dihedral angles becomes larger due to the additional degrees of freedom introduced by the additional glycosidic bonds.

Figure 4 is a map showing which regions of the oligosaccharide conformational space, determined by the glycosidic dihedral angles Φ and Ψ, give simulated NOE agreeing within experimental error with the NOESY data. The three panels describe the conformation about the three glycosidic linkages in the tetrasaccharide model of the Lewis^b determinant of LND-1. (See Table 1.) Although each panel shows a single glycosidic linkage, the entire oligosaccharide conformation is included at each point in the plane allowing us to describe in two dimensions what is actually a multidimensional problem.[37]

The NOE maps in Figure 4 a–c are obtained from a simulation at 500 MHz, 250 ms mixing time in D_2O, 5°C for LND-1. The regions within the loops are those in which the calculated and observed ratio of the NOE between protons directly across the glycosidic bond to an intra-ring NOE agree within error bounds estimated from the experimental data. The calculated NOE's in this case were obtained using a disaccharide model for each linkage. Results of the second stage in the simulation using a tetrasaccharide model to simulate all the experimental NOE's are shown by the heavy dots within the loops. The tetrasaccharide conformations along each glycosidic linkage have been plotted separately, but are actually dependent on the conformations of the other two linkages, as explained in the previous paragraph. The regions in which there is agreement within the estimated experimental error limits between observed and calculated nonselective T_1 values are shown by the points marked by ×.

For oligosaccharide fragments of up to four residues, such as the tetrasaccharide fragment represented by the data of Figure 4, an exhaustive search on the dihedral angles is practical. For more than 6 or 8 degrees of freedom this procedure is impossible because the computer time required for an exhaustive search increases exponentially with increasing numbers of dihedral angles. Therefore this approach cannot be used for proteins nor would it be useful for small peptides since they generally do not assume single rigid conformations and would be expected to fail the test of the rigidity assumed in our model.

For oligosaccharides, we consider three possible cases, in the first of which no single conformation gives NOE results in agreement with experiment. Such a result implies that the oligosaccharide does not adopt any single conformation and that there must be conformational averaging. A second possibility is that agreement between the experiment and the simulation occurs for just one conformation or a small group of closely related conformations. It is that case which is illustrated in the results of Figure 4 for the Lewis[b] tetrasaccharide determinant in LND-1.[37] If the single conformation occurs at a low energy then the most direct interpretation is that the oligosaccharide exists in a single conformation. If there are several conformations which agree with the experiment, and if only one among them is at low energy, one can reasonably conclude that to be the single conformation. For all the blood group determinants in Table 1, results of the type described above have pointed to a single conformation.[29,37,39] While the data do not rule out conformational averaging, they do provide a strong argument that the blood group determinants exist in a single conformation. One should then seek methods to further test the hypothesis of a single conformation to verify or disprove it.

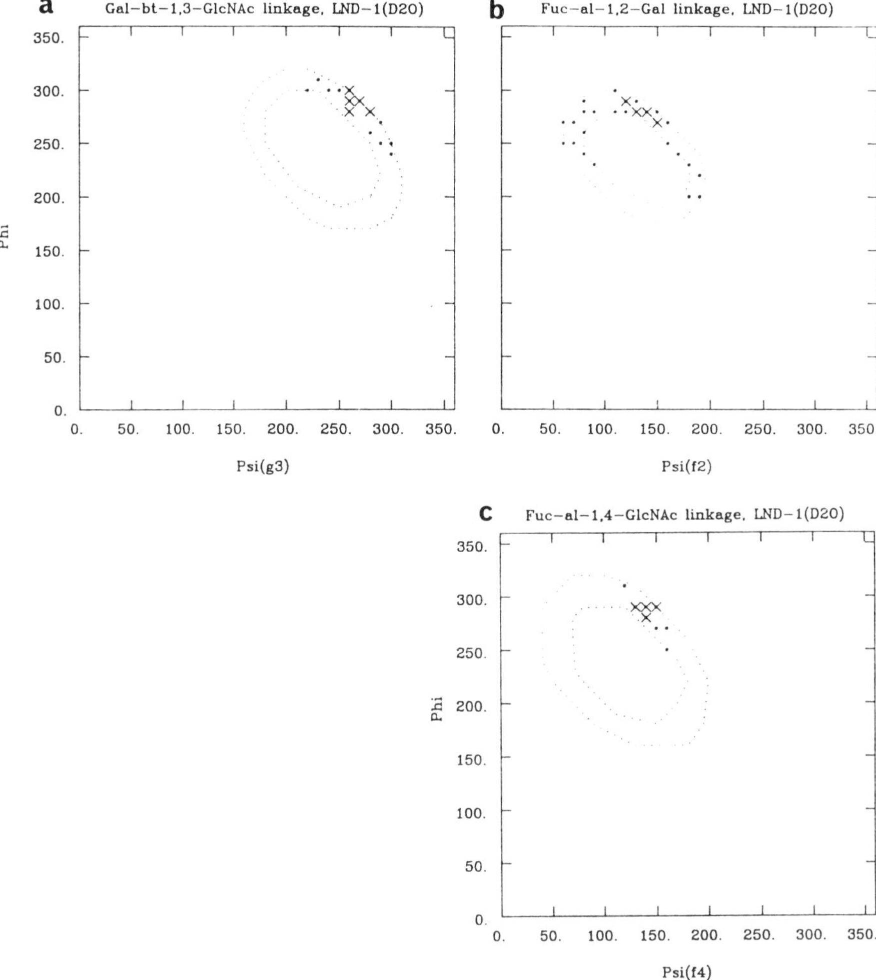

Figure 4. (a₁–a₃) Maps of simulated NOEs (500 MHz, 250 ms mixing time, 5°, D₂O) as a function of conformation of each glycosidic linkage in the non-reducing end of LND-1. Results using disaccharide models for each linkage are shown by the regions within the loops, within which calculated and observed values agree for: (a) the ratio of NOE between Gal³ H1/GlcNAc H3 to Gal³ H1/H3; (b) ratio of NOE between Fuc² H1/Gal³ H2 to that between Fuc² H1/H2; and (c) ratio of NOE between Fuc⁴ H1/GlcNAc H4 to that between Fuc⁴ H1/H2. Simulation of NOE ratios of all three interglycosidic NOEs and two remote NOEs using a tetrasaccharide model yielded conformations for each linkage illustrated by the dots within the loops, while the overlapping x-marked points are those in which calculated and observed nonselective T_1 values agreed. Note that each dot shown above was obtained by plotting only that particular linkage in the tetrasaccharide without regard to the conformations of the other two linkages, i.e., each pair of angles for a particular linkage corresponds to several conformations for the other two linkages. Taken from Ref. 37 with permission.

IV. ENERGY CALCULATIONS IN OLIGOSACCHARIDES

A. Rigid Geometry Methods

In parallel with the NOE simulations discussed above, energy calculations were performed to which we have referred above. In these simulations, the pyranoside rings were held rigid in the chair conformations indicated by the $^3J_{HH}$ and only the glycosidic dihedral angles Φ and Ψ were varied using an approach called the 'rigid geometry' method. The classical potential energy functions used involve non-bonded interactions, electrostatic effects, hydrogen bonding and bond torsional potentials. The energy calculations, based on exhaustive searches of the glycosidic angle dihedral space, showed that the predominant influence in determining the low energy conformation was steric repulsion.[37,39] If the conformation of an oligosaccharide is determined mainly by steric interactions, or more exactly by the repulsive part of the non-bonded interactions, we expect the potential energy barriers surrounding any energy minimum to be high on account of the steeply rising repulsive energy. Therefore some explicit predictions about the dependence of conformation on temperature and on the nature of the solvent medium are possible. If high energy barriers surrounding any potential energy minimum are much greater than kT, changes of temperature in the range of 0 to 100 °C are not be expected to cause a change in oligosaccharide conformation. This prediction contrasts with the case of peptides and polynucleotides many of whose conformations are stabilized by specific low energy bonds such as hydrogen bonds having an enthalpy of stabilization of the order of 2–5 kcal/mole. Although thermal denaturation phenomena are well known for proteins and polynucleotides, they have been shown to be absent in the temperature dependence of CD and of NOE of blood group oligosaccharides.[29,39]

The predominant influence of non-bonded interactions in oligosaccharide conformations has strong implications about the dependence of conformation on the solvent medium. Since hydrophobic forces depend explicitly on the structure of the water and hydrogen bonds are mainly electrostatic, the strength of both types of interaction depends strongly on solvent structure and dielectric constant. In contrast, non-bonded interactions or torsional potentials should not depend so strongly on solvent and we predict that conformations similar to those found in water might be found for blood group oligosaccharrides in non-aqueous solvents. This prediction has been tested in NOE studies of blood group oligosaccharides which compared the conformations in aqueous solution, mixtures of Me_2SO/water and pyridine solution. Although the relative chemical shifts of the protons of the blood group oligosaccharides differ greatly among the solvents D_2O, Me_2SO and pyridine, we deduce from the similar ratios of NOE that the conformations are not greatly different.[36,37,45]

B. Methods Involving All Degrees of Freedom

The rigid geometry method of energy calculation has been used by many workers in carbohydrate chemistry with slightly different energy parameters and the results of these calculations can depend somewhat on the choice of empirical energy parameters.[39] The most widely applied of the rigid geometry parameter sets is the HSEA method of Lemieux and Bock.[33] The primary advantage of rigid geometry methods comes from their use of internal coordinates, in this case the glycosidic dihedral angles, Φ and Ψ. This reduces the number of degrees of freedom to a minimum making possible exhaustive searches of the oligosaccharide conformational space at least for oligosaccharides of modest size for which the number of dihedral angles can be held to six or eight. For larger oligosaccharides requiring the use of energy minimization rather than an exhaustive search, rigid geometry methods suffer from trapping in local minima.

The main weakness of the rigid geometry method is that no bond angles can deviate from fixed values, nor can any of the ring dihedral angles relax to accommodate steric clashes. Neither can the dihedral angles describing the positions of the exocyclic hydroxyl and methoxyl groups change. Therefore, the method yields unrealistically high barriers to rotation causing serious artifacts in energy minimization schemes resulting from trapping in local minima. Steric conflicts of the exocyclic groups can generally be resolved by simple inspection of the model but other barriers can be more subtle and failure to relax bond angles and ring dihedral angles can give results which are quite misleading. It is possible to vary all degrees of freedom of the oligosaccharide including bond lengths, bond angles and all the dihedral angles. But the increased number of degrees of freedom renders it impossible to carry out an exhaustive search. It is our opinion that by a judicious combination of exhaustive searches with rigid geometry models and careful use of the methods in which all degrees of freedom are relaxed, it is possible to avoid the pitfalls of each method.

The earliest schemes for searching conformational spaces generally involved minimization of the total energy following any of a number of mathematical algorithms. But an especially efficient way to search conformational space, which has been recently applied to carbohydrates, is the method of molecular dynamics simulation in which Newton's equations of motion are numerically solved for the classical potential of the molecule.[46] Not only does this method provide an efficient way to search the conformational space of the oligosaccharide, but it can also provide limited information on the kinetics of conformational exchange. Although the blood group oligosaccharides discussed above provide an example of relatively rigid conformations, we will see below that there is strong evidence for flexibility in some other classes of oligosaccharides. As will be discussed below, understanding the internal motion in such oligosaccharides presents a formidable problem to which both MD simulation and NMR spectroscopy can make a contribution.

In MD simulations of some of the fragments making up the blood group

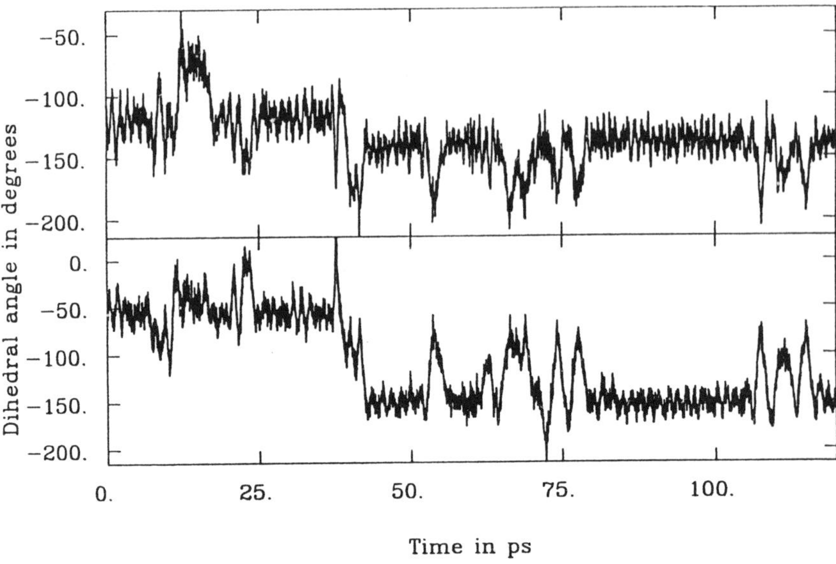

Figure 5. Histories of the glycosidic dihedral angles Φ (lower trace) and Ψ (upper trace) for Gal β-(1→3)GlcNAc β-O-Me, taken from Ref. 48 with permission.

determinants whose structures are given in Table 1, trajectories up to a few hundred picoseconds in length have been reported. These simulations show that the sugar rings undergo small fluctuations about the equilibrium chair conformation, a result which is consistent with the interpretation of the vicinal proton coupling constants, $^3J_{HH}$ as discussed above.[46,47] The question of conformational exchange of the oligosaccharide can be addressed through the history of the glycosidic dihedral angles Φ and Ψ. Fluctuations of these dihedral angles indicate conformational transitions.

Figure 5 illustrates the behavior of the glycosidic dihedral angles Φ and Ψ for the disaccharide, gal-β-(1→3)glcNAc-β-O-me, a fragment which models a portion of the type 1 chain found in a number of blood group oligosaccharides. (See Table 1.) The data of Figure 5 show small fluctuations of ±20° in the dihedral angles on a subpicosecond time scale. In addition there are larger motions which are correlated between the angles Φ and Ψ which represent conformational transitions, in some cases leading to new energy minima corresponding to distinct conformations.[48] Although this disaccharide shows a number of conformational transitions, some other disaccharides, such as fuc- α-(1→2)gal- β-O-me and gal- β-(1→4)glcNAc show fewer transitions in similar trajectories suggesting that they

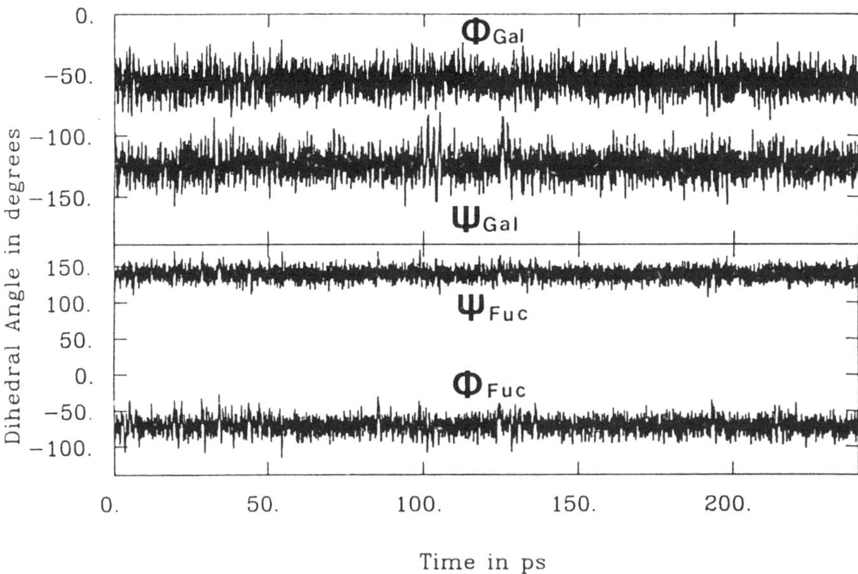

Figure 6. Histories of the glycosidic dihedral angles for Fuc α-(1→2)Gal β-(1→3)GlcNAc β-O-Me over 240 ps, taken from Ref. 48 with permission.

may adopt more rigid conformations. The issue of flexibility in disaccharides will be dealt with in more detail below.

One of the more interesting results of the work of Yan and Bush[48] is that when larger blood group oligosaccharide fragments are generated by combining two disaccharides, the amount of internal motion is greatly reduced. Figure 6 shows the histories of the same dihedral angles as Figure 5 in addition to the history of the dihedral angles of the linkage fuc- α-(1→2)gal in the Lewis^d trisaccharide fragment fuc- α-(1→2)gal- β-(1→3)glcNAc. This trajectory shows that extension of the disaccharide to the linear trisaccharide freezes out the conformational transitions which occurred in the disaccharide (Figure 5). The trisaccharide features new steric interactions which severely limit the number of available conformations. These interactions are evident in the stereopair model of Figure 7, based on experimental NMR data, which shows the tightly folded conformation of this trisaccharide fragment in the milk pentasaccharide, lacto-N-fucopentaose 1 (LNF-1).[29,37] The model (Figure 7) shows that the fucosyl residue folds back and interacts with the glcNAc residue, a feature which is supported by the experimental data. These MD results are consistent with the conclusions of the NMR and rigid geometry modeling calculations on blood group oligosaccharides discussed above.

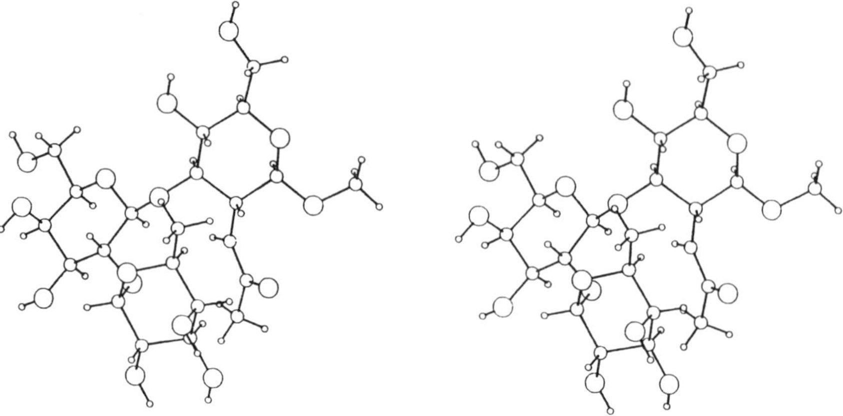

Figure 7. Stereopair model of the Fuc α-(1→2)Gal β-(1→3)GlcNAc β-O-Me (LNF-1) trisaccharide, based on data from Refs. 29 and 37.

V. EVIDENCE FOR FLEXIBILITY IN COMPLEX OLIGOSACCHARIDES

A. Experimental NOE Data on Glycopeptides

Although the evidence for rigidity of blood group oligosaccharides is reasonably convincing, one should not be seduced into the error, so common in the biological sciences, of extrapolating from one case generalizing this result to all polysaccharides. Many complex carbohydrates contain linkages which are obvious candidates for flexibility. Most obvious are internal (1→6) linkages which have three dihedral angles to be fixed and a greater distance between the linked residues. A number of other linkages could be flexible as well. Both experimental evidence and computational modeling on different types of sugars provide some evidence for internal motion. We propose that since there may be some biological significance to internal motion in complex oligosaccharides, it could be valuable to develop criteria for distinguishing rigid from flexible structures and to identify those points where flexibility exists in larger oligosaccharides. In an initial effort toward this goal, we will review the evidence from NMR studies for internal motion in oligosaccharides.

The first attempt at a rigorous interpretation of NOE data on complex oligosaccharides appeared in an important series of papers by Brisson and Carver[26–28] treating the mannose-containing oligosaccharides which are found in the *N*-linked glycopeptides of glycoproteins. In these papers, Brisson and Carver assumed rigid conformations and developed a quantitative interpretation of their steady-state

NOE data by a matrix method. In spite of their initial conclusion that the *N*-linked glycopeptides adopt well-defined single conformations, Carver and coworkers later concluded that this interpretation could not survive more careful scrutiny. They tested the proposed conformations of the man-α-(1→3)man and of the man-α-(1→6) man linkage by study of carefully designed model compounds and concluded that conformational averaging must be invoked to explain the experimental data on the particularly important man- α-(1→6) [man- α-(1→3)] man β-branching point in these glycopeptides.[49-51] In the most definitive experiments, they have used synthetic mannose oligosaccharides containing deuterium atoms strategically placed so as to simplify the interpretation of the NOE data.[52] More recently these workers have reviewed and summarized the data showing that it is not possible to reconcile any single conformation of these two linkages with experimental NOE data.[53] Preliminary MD simulations on the man- α-(1→3)man disaccharide also provide clear evidence for multiple conformational minima with transitions among them.[54] Since the methods used in these studies by Carver and coworkers do not differ in any fundamental way from those used in the work presented here on the blood group oligosaccharides, the conclusion is that there are real differences in the conformational flexibility of blood group oligosaccharides and the trimannosyl core of *N*-linked glycopeptides.

For the quantitative interpretation of NOE in flexible oligosaccharides, Carver and coworkers have treated the multiple conformation problem by introduction of the concept of "virtual conformations."[50] Energy calculations are carried out with an exhaustive search method and a statistical weight assigned from a Boltzmann weighting factor following standard methods of statistical thermodynamics. In order to calculate the NOE for comparison with experiment, the cross relaxation rates of each geometric model are weighted by a Boltzmann factor and the statistical average relaxation rate matrix is constructed. This procedure assumes that the conformations exchange at a rate which is slow compared to the rotational correlation time of the isotropically tumbling oligosaccharide (typically $\sim$ 1 ns) but fast compared to T_1 (typically $\sim$ 100 ms).[38] While this may seem to be a reasonable assumption, very little data exist to support it and the MD simulations suggest that the conformational exchange might be too rapid to be consistent with this model. While the detailed kinetics of MD simulations such as those represented in Figure 5 should not be taken as definitive, nevertheless it should be clear that the conformational transitions in this typical trajectory occur on a time scale comparable to or slightly faster than the rotational correlation time. Thus it is not inconceivable that internal motion in disaccharides might contribute to T_1. Oligosaccharide flexibility introduces some serious complications to the interpretation of NOE and more information will be needed before we can be confident of just which approximations are suitable.

A group at Oxford has also studied these same *N*-linked glycopeptides by NMR methods and molecular modeling and, using different assumptions, have reached conclusions which differ somewhat from those of Carver and coworkers.[55,56]

Originally there was some disagreement on the preferred conformation of the man-α-(1→3)man linkage and there continues to be some difference of opinion regarding the influence of solvent on oligosaccharide flexibility.[57,58] However, there is a consensus that the conformation of N-linked glycopeptides cannot be represented by any one single conformational model.

Glaudemans et al.[59] have reported NMR data relevant to the conformation of the disaccharide gal- β-(1→6) gal in solution including $^3J_{HH}$ for determination of the C5—C6 bond dihedral angle ω as well as ^{1}H NOE data including a detailed fit of the data to models. They found no single model which could explain all the data implying flexibility of the disaccharide. They further measured transferred NOE on binding the disaccharide to a specific monoclonal antibody and showed that the distribution of conformers changes on binding providing further support for the proposal that internal motion exists for this disaccharide. This conclusion should not be too surprising in a disaccharide with a (1→6) linkage which should generally be expected to be flexible.

Lipkind et al.[60] report data on ^{1}H NOE, on the ^{13}C—^{1}H coupling constants and the optical rotation for cellobiose along with conformational calculations. Energy calculations based on a rigid geometry model give four different energy minima, one of which corresponds to the structure found by X-ray crystallography. But the proton NOE cannot be interpreted with any single conformation and they employ a weighted average of the conformations to calculate the experimental parameters. They conclude that two distinct conformers contribute most of the weight with two additional conformations which must be included to explain the NOE results.

B. Some Results on Glycolipids

Although this review has not, to this point, treated glycolipids, these important components of the plasma membrane of eucaryotic cells carry oligosaccharide chains having some similarities to those of glycoproteins and bacterial polysaccharides. Glycolipids are amphiphilic molecules with a highly hydrophobic sphingolipid domain covalently connected to a highly polar carbohydrate chain. In nature the lipid is inserted in the plasma membrane and the oligosaccharide is exposed to the aqueous surroundings of the cell. Due to their amphipathic nature, isolated glycolipids in aqueous solution aggregate to form micelles which, due to long rotational correlation times do not lend themselves to study by high resolution NMR spectroscopy. But in Me$_2$SO solution, glycolipids are generally not aggregated and give reasonably high resolution ^{1}H NMR spectra.

Prestegard and coworkers[61] have studied the gangliosides GM1, GM1b and GD1a in Me$_2$SO solution by ^{1}H NOESY methods in combination with molecular modeling. They measured NOESY at a single mixing time chosen to be in the region of linear buildup so the data could be semi-quantitatively interpreted with the isolated spin pair approximation (ISPA). Molecular modeling was carried out with all degrees of freedom relaxed using an energy minimization method which was restrained by the distances deduced from the ISPA analysis of the NOESY data. All three glycolipids share a common tetrasaccharide core, gal- β-(1→3)galNAc- β-(1→4)gal- β-(1→4)glc- β-with NeuAc- α-(2→3)substituents at the terminal and at the internal galactose. Their data show that the core tetrasaccharide adopts different conformations depending on the positions of substitution by NeuAc thus proving that the core must have some flexibility. Hydrogen bonding is proposed between the carboxyl group of NeuAc and the galNAc amide and between the two NeuAc residues of the GD1a.

Also working in glycolipids, Dabrowski and coworkers have studied Me_2SO solutions of globotriaosyl ceramide and isoglobotriaosyl ceramide, the carbohydrate chains of which are gal- α-(1→4)gal- β-(1→4)glc- β-, and gal- α-(1→3)gal- β-(1→4)glc- β-.[62] They report NOESY as well as rotating frame ROESY data including the signals of the hydroxyl protons. Their NOE results were interpreted with an ISPA method with the distances used as constraints in the molecular modeling. They conclude that both oligosaccharides exist in multiple conformations which interconvert on a time scale faster than milliseconds. The conformation of the isolated oligosaccharide of globotriaosyl ceramide dissolved in D_2O appeared to be the same as that of the corresponding part of the glycosphingolipid in Me_2SO solution.

While the examples of experimental data quoted above all provide good evidence for internal motion in various complex oligosaccharides, none of the structures from either glycolipids or glycoproteins discussed above includes a blood group determinant. A survey of literature on the conformation of blood group oligosaccharides provides mixed evidence for flexibility. The original data of Lemieux et al.,[33] although interpreted as supporting rigid models, are unconvincing since the method used in this work started with the premise that the minimum energy conformation determined from the rigid geometry HSEA model was the only contributing conformation. Although qualitative NOE data were reported, the data did not formally test the model. Bechtel et al.[63] have reported NOESY data on the Lewis[a] Oligosaccharide, the structure of which is included in the milk oligosaccharide, LNF-2 (Table 1). Using a very long mixing time, they attempted to apply the ISPA method concluding that their failure to find a single conformation compatible with the NOE data implied internal motion. Their interpretation has been criticized by Cagas and Bush[37] who reported NOESY and T_1 data which were interpreted by a full spin matrix analysis leading to the conclusion that a single low energy conformation could indeed explain the data. It is our conclusion that the data of Bechtel et al.[63] can be explained by three-spin effects augmented by spin diffusion which would be anticipated at the long mixing times used in their experiments.

C. Evidence from Molecular Modeling for Flexibility

One criticism of the rigid geometry molecular modeling studies used in the interpretation of some of the NOE data is that this method builds in unrealistically high barriers to internal motion, thus guaranteeing that rigid models will be deduced for oligosaccharides. Therefore, methods in which all degrees of freedom can be varied are preferred for critical molecular modeling studies. Molecular dynamics simulations have provided the most convincing evidence for flexibility.[32] Ha et al.[64] have reported a careful study of the potential energy surface for maltose identifying several distinct conformational minima on the relaxed surface. Molecular dynamics simulations illustrated conformational transitions among these energy minima. Carver et al.[54] have reported MD simulations on man $(1{\rightarrow}3)$man and have also found conformational transitions among several distinct energy minima. The trajectory illustrated in Figure 5 illustrates that, although the emphasis in our MD simulations of blood group oligosaccharides was on the rigidity of the structures, numerous conformational transitions were found in some of the component disaccharides which make up the blood group substances.[48] The field of molecular dynamics simulations of complex carbohydrates is in a very early stage of development and many fundamental questions remain unanswered. Probably the two most important uncertainties are the choice of force field parameters and the importance of inclusion of solvent in the simulation. But, one can conclude that there is good evidence from both experimental NMR and molecular modeling for flexibility of certain oligosaccharides.

It is our interpretation that the evidence for rigidity in blood group oligosaccharides can be reconciled with evidence for flexibility in other oligosaccharides. We believe that certain stereochemical features of the blood group oligosaccharides severely restrict the conformational space available to the sugars. It is our interpretation that the ensemble theory of Cumming and Carver[49-51] is essentially correct and it can be applied to both flexible and rigid oligosaccharides. In this notation, the oligosaccharides which are rigid simply include a rather narrow distribution of conformations in the partition function and it is dominated by a few terms representing very similar conformations. Comparatively more flexible oligosaccharides have a broader distribution of contributing conformations. In statistical thermodynamic terms, one could characterize the flexibility of an oligosaccharide by its absolute entropy.

D. Possible Tests for Oligosaccharide Flexibility

This interpretation suggests that it will be important to devise tests to discriminate between relatively rigid and flexible conformations to determine whether there might be some biological significance to the flexibility of complex oligosaccharides as has been suggested by Carver et al.[53] Simple ^{1}H NMR spectra of larger oligosaccharides and polysaccharides illustrate that there must be some internal motion. Relatively narrow lines can be observed for polysaccharides with 100,000 Da molecular weight. Globular proteins of this molecular size generally have slow rotational correlation times, short values of T_2 and linewidths too great to allow for detection of spin coupling of 5 to 10 Hz such as we see in the spectra of Figures 1 and 2 above. What is the origin of this motion in complex carbohydrates? Are there "hinges" in oligosaccharide structures which introduce the flexibility? Such flexibility in the branch points of the N-linked glycopeptides could be very relevant to lectin binding of multivalent oligosaccharides.

One experimental test for rigidity in oligosaccharides already described above consists of comparison of NOE data with simulations from model conformations. If a single low energy conformation which fits the NOE data can be found, this may be taken as evidence of rigidity. This criterion has been criticized because the strong dependance of NOE on inter proton distance introduces a heavy bias in favor of conformations having protons at a short distance. Three-bond coupling constants, in particular $^3J_{CH}$, which can be correlated with glycosidic dihedral angles, offer a useful ancillary measure since their sensitivity to conformational averaging differs greatly from that of NOE.[31] Unfortunately measurement of $^3J_{CH}$ is technically difficult and little data are now available for complex oligosaccharides. Although ^{13}C T_1 can be used as a direct measure of internal motion in oligosaccharides, the rotational correlation times, τ_c, of oligosaccharides of modest size fall in a range which lies at a minimum of T_1 versus τ_c curve and T_1 is not very sensitive to internal motion. This fact diminishes the value of existing ^{13}C T_1 data on milk oligosaccharides at single spectrometer frequency which show that all the T_1 for the methine carbons of a series of tetra and pentasaccharides are similar as would be expected for a rigid oligosaccharide.[65] ^{13}C T_1 data reported for pig gastric mucin show that the blood group A determinant in these glycoproteins is relatively rigid.[66] ^{13}C T_1 recorded as a function of magnetic field strength provides a more critical test for internal motion allowing for analysis of the dynamics by a model-free approach. Although such measurements have not been reported for any complex oligosaccharides, McCain and Markely have reported extensive ^{13}C T_1 data for sucrose accompanied by a rigorous analysis.[67] They concluded that the sucrose ^{13}C T_1 data are consistent with a reasonably rigid molecule which is consistent with the molecular dynamics simulations described above. Application of such methods to larger complex carbohydrates such as blood group oligosaccharides or N-linked glycopeptides would provide valuable evidence on the question of internal motions on a time scale of the reciprocal of NMR frequency range used in the experiment (0.1 to 1 ns). Motions on a slower time scale, which could not be detected by ordinary T_1 experiments, could be studied by rotating frame relaxation data, $T_{1\rho}$. With control of the spin-locking field strength, this experiment can be sensitive to motions on the microsecond time scale. Ultrasonic absorption measurements could also be valuable in detecting motions in this same time range. It is unlikely that motions occurring on a slower time scale, for example, milliseconds or slower, are important in complex oligosaccharides as conformational exchange on such a slow time scale could be detected in the linewidths in oligosaccharide spectra.

VI. THE FUTURE

Knowledge of the structure and conformation of complex oligosaccharides will contribute to our understanding of their biological function and to our ability to modify that function by therapeutic measures such as drug design. It seems likely this will require investigations of the interactions of the oligosaccharides with the lectins responsible for decoding information they must contain regarding cell interactions, growth and differentiation. Ultimately these interactions must be studied in such animal systems as hepatic lectin or soluble lung lectins which are directly involved in the biologically significant functions.[6] For the present, however, the animal lectins, which have been only recently discovered, are not readily available for study and the plant lectins, which are available as a surrogate system, have no known biological function. Current work on the interaction of complex carbohydrates with plant lectins indicates very complex thermodynamics of binding with a number of relevant simultaneous equilibria.

In this review we have emphasized, not the problem of lectin binding, but simply the problem of conformation of their complex carbohydrate receptors. We propose that knowledge of just which complex oligosaccharides are rigid and which are flexible as well as the factors such as "hinges" which govern their flexibility will be an important prerequisite to quantitative study of the more relevant problem of lectin binding. While we have not proposed a complete solution to this problem, we believe that methods capable of solving it are available. NMR, including NOE and vicinal coupling constants, combined with molecular modeling and X-ray crystallography can provide answers to the questions we have posed in this review.

ACKNOWLEDGMENTS

Research was supported by NIH grants GM–31449 and DE–09445.

REFERENCES AND NOTES

1. Kornfeld, R.; Kornfeld, S. *Annual Reviews Biochem.* **1985**, *54*, 631–664.
2. Lodish, H. F. *J. Biol. Chem.* **1988**, *263*, 2107–10.
3. Yet, M. G.; Shao, M.-C.; Wold, F. *The FASEB J.* **1988**, *2*, 22–31.
4. Reading, C. L.; Hutchins, J. T. *Cancer Metastasis Rev.* **1985**, *4*, 221–60.
5. Phillips, M. L.; Nudelman, E.; Gaeta, C. A.; Perez, M.; Singhai, A. K.; Hakomori, S. I.; Paulson, J. C. *Science* **1990**, *250*, 1130–32.
6. Drichamer, K. *J. Biol. Chem.* **1988**, *263*, 9557–60.
7. Leffler, H.; Masiarz, F. R.; Barondes, S. H. *Biochemistry* **1989**, *28*, 9222–29.
8. Lee, R. T.; Rice, K. G.; Rao, N. B. N.; Ichikawa, Y.; Barthel, T.; Piskarev, V.; Lee, Y. C. *Biochemistry* **1989**, *28*, 8351–58.
9. Bhattacharyya, L.; Fant, J.; Lonn, H.; Brewer, C. F. *Biochemistry* **1990**, *29*, 7523–30.
10. Leffler, H.; Svanborg-Eden C. *Microbial Lectins and Agglutinins: Properties and Biological Activity*; Mirelman, D., Ed.; Wiley; New York, 1986.

11. Karlsson, K. A. *Pure and Appl. Chem.,* **1987**, *59*, 1477–88.

12. Clore, G. M.; Kay, L. E.; Bax, A.; Gronenborn, A. M. *Biochemistry* **1991**, *30*, 12–18.

13. Stockman, B. J.; Markley, J. L. In *Advances in Biophysical Chemistry*, Vol. 1; Bush, C. A., Ed.; JAI Press: Greenwich, CT; 1990.

14. Delbaere, L. T. J.; Vandonselaar, M.; Prasad, L.; Quail, J. W.; Peralstone, J. R.; Carpenter, M. F.; Spohr, U.; Lemieux, R. U. *Canad. J. Chem.* **1990**, *68*, 1116–21.

15. Bourne, Y.; Rouge, P.; Cambillau, C. *J. Biol. Chem.* **1990**, *265*, 18161–65.

16. Sharon, N.; Lis, H. *The FASEB J.* **1990**, *4*, 3198–3208.

17. Fesik, S. W.; Zuiderweg, E. R. P. *Quarterly Reviews Biophysics* **1990**, *23*, 97–131.

18. Wuthrich, K. *J. Biol. Chem.* **1990**, *265*, 22059–62.

19. Clore, G. M.; Gronenborn, A. M. *J. Magn. Reson.* **1989**, *84*, 398–409.

20. Bax, A. *Annual Rev. Biochem.* **1989**, *58*, 223–256.

21. Dua, V. K.; Rao, B. N. N.; Wu, S. S.; Dube, V. E.; Bush, C. A. *J. Biol. Chem.* **1986**, *261*, 1599–1608.

22. Bush, C. A. *Bull. Magn. Reson.* **1988**, *10*, 73–95.

23. Abeygunawardana, C.; Bush, C. A.; Cisar, J. O. *Biochemistry* **1990**, *29*, 234–248.

24. Abeygunawardana, C.; Bush, C. A.; Cisar, J. O. *Biochemistry,* **1991**, *30*, 6528–6540.

25. deWaard, P.; Boelens, R.; Vuister, G. W.; Vliegenthart, J. F. G. *J. Am. Chem. Soc.* **1990**, *112*, 3232–34.

26. Brisson, J.-R.; Carver, J. P. *Biochemistry* **1983**, *22*, 1362–68.

27. Brisson, J.-R.; Carver, J. P. *Biochemistry* **1983**, *22*, 3671–3660.

28. Brisson, J.-R.; Carver, J. P. *Biochemistry* **1983**, *22*, 3680–3686.

29. Rao, B. D. N.; Dua, V. K.; Bush, C. A. *Biopolymers* **1985**, *24*, 2207–29.

30. Morat, C.; Taravel, F. R. *Tetrahedron Lett.* **1988**, *29*, 199–200.

31. Tvaroska, I.; Hricovini, M.; Petrakova, E. *Carbohydr. Res.,* **1989**, *189*, 359–62.

32. Brady, J. W. In *Advances in Biophysical Chemistry*, Vol. 1; Bush, C. A., Ed.; JAI Press: Greenwich, CT, 1990.

33. Lemieux, R. U.; Bock, K.; Delbaere, L. T. J.; Koto, S.; Rao, V. S. R. *Can. J. Chem.* **1980**, *58*, 631–53.

34. Venkatachalam, C. M.; Ramachandran, G. N. In *Conformation of Biopolymers 2*, Ramachandran, G. N., Ed.; Academic: New York, 1967.

35. Sundarajan, P. R.; Rao, V. S. R. *Tetrahedron* **1968**, *24*, 289–95.

36. Rao, B. N. N.; Bush, C. A. *Carbohydr. Res.* **1988**, *180*, 111–28.

37. Cagas, P.; Bush, C. A. *Biopolymers* **1990**, *30*, 1123–38.

38. Noggle, J. H.; Schirmer, R. E. *The Nuclear Overhauser Effect;* Academic Press: New York, 1971.

39. Bush, C. A.; Yan, Z.-Y.; Rao, B. N. N. *J. Am. Chem. Soc.* **1986**, *108*, 6168–73.

40. Keepers, J. W.; James, T. L. *J. Magn. Reson.* **1984**, *57*, 404–26.

41. Kumar, A.; Wagner, G.; Ernst, R. R.; Wuthrich, K. *J. Am. Chem. Soc.* **1981**, *103*, 3654–58.

42. Mirau, P. *J. Magn. Reson.* **1988**, *80*, 439–47.

43. Mirau, P.; Bovey, F. *J. Am. Chem. Soc.* **1986**, *108*, 5130–34.

44. Summers, M. F.; South, T. L.; Kim, B.; Hare, D. R. *Biochemistry* **1990**, *29*, 329–40.

45. Yan, Z.-Y.; Rao, B. N. N.; Bush, C. A. *J. Am. Chem. Soc.* **1987**, *109*, 7663–69.

46. Brady, J. W. *J. Am. Chem. Soc.* **1986**, *108*, 8153–60.

47. Brady, J. W. *Carbohydr. Res.* **1987**, *165*, 306–12.

48. Yan, Z.-Y.; Bush, C. A. *Biopolymers* **1990**, *29*, 799–812.

49. Cumming, D. A.; Shah, R. N.; Krepinsky, J. J.; Grey, A. A.; Carver, J. P. *Biochemistry* **1987**, *26*, 6655–63.

50. Cumming, D. A.; Carver, J. P. *Biochemistry* **1987**, *26*, 6664–76.

51. Cumming, D. A.; Carver, J. P. *Biochemistry* **1987**, *26*, 6676–83.

52. Cumming, D. A.; Grey, A..; Krepinsky, J. J.; Carver, J. P. *J. Biol. Chem.* **1986**, *261*, 3208–13.

53. Carver, J. P.; Michnick, S. W.; Imberty, A.; Cumming, D. A. In *Glycoprotein and Biological Recognition* CIBA Foundation Symposium, Harnett, S., Ed.; Wiley: London, 1989.

54. Carver, J. P.; Mandel, D.; Michnick, S. W.; Imberty, A.; Brady, J. W. In *Computer Modeling of Carbohydrate Molecules* French, A. D.; Brady, J. W., Eds.; ACS Symposium Series 430; American Chemical Society: Washington, DC, 1990.

55. Homans, S. W.; Dwek, R. A.; Rademacher, T. W. *Biochemistry* **1987**, *26*, 6553–60.

56. Homans, S. W.; Dwek, R. A.; Rademacher, T. W. *Biochemistry* **1987**, *26*, 6571–8.

57. Edge, C. J.; Singh, U. C.; Bazzo, R.; Taylor, G. L.; Dwek, R. A.; Rademacher, T. W. *Biochemistry* **1990**, *29*, 1971–74.

58. Homans, S. W. *Biochemistry* **1990**, *29*, 9118–19.

59. Glaudemans, C. P. J.; Lerner, L.; Daves, G. D.; Kovac, P.; Venable, R.; Bax, A. *Biochemistry* **1990**, *29*, 10906–11.

60. Lipkind, G. M.; Shashkov, A. S.; Kochetkov, N. K. *Carbohydr. Res.* **1985**, *141*, 191–7.

61. Scarsdale, J. N.; Prestegard, J. H.; Yu, R. K. *Biochemistry* **1990**, *29*, 9843–55.

62. Poppe, L.; Dabrowski, J.; Von de Lieth, C. W.; Koike, K.; Ogawa, T. *Eur. J. Biochem.* **1990**, *189*, 313–25.

63. Bechtel, B.; Wand, A. J.; Wroblewski, K.; Koprowski, H.; Thurin, J. *J. Biol. Chem.* **1990**, *265*, 2028–37.

64. Ha, S. N.; Madsen, L. J.; Brady, J. W. *Biopolymers* **1988**, *27*, 1927–52.

65. Bush, C. A.; Panitch, M.; Dua, V. K.; Rohr, T. E. *Anal. Biochem.* **1985**, *145*, 124–36.

66. Gerken, T. A.; Jentoft, N. *Biochemistry* **1987**, *26*, 4689–99.

67. McCain, D. C.; Markley, J. L. *J. Am. Chem. Soc.* **1986**, *108*, 4259–64.

68. The glycosidic dihedral angle Φ is defined by the four atoms, O_{ring}-C1-O-C_x, and Ψ is defined by C1-O-C_x-C_{x-1}, according to the IUPAC convention which is preferred over other definitions based on hydrogen atoms.